Materials Modelling: From Theory to Technology

Dr Ron Bullough FRS in the grounds of St Edmund Hall, Oxford, at the time of the Symposium on Materials Modelling: From Theory to Technology.

Materials Modelling:
From Theory to Technology

Proceedings of a symposium
held in honour of the 60th birthday of
Dr Ron Bullough FRS
St Edmund Hall, Oxford
26th & 27th September 1991

Edited by:

C A English
J R Matthews
H Rauh†
A M Stoneham FRS
R Thetford

AEA Technology, Harwell Laboratory
†Also Department of Materials and
Wolfson College, University of Oxford

The Symposium was sponsored by AEA Technology

Institute of Physics Publishing
Bristol and Philadelphia

British Library Cataloguing in Publication Data

A catalogue record for this book is available from the British Library.

ISBN 0–7503–0196–1

Library of Congress Cataloging-in-Publication Data are available

Published by IOP Publishing Ltd, a company wholly owned by
The Institute of Physics, London.
Techno House, Redcliffe Way, Bristol BS1 6NX, England
US Editorial Office: IOP Publishing Inc., The Public Leger Buildings, Suite 1035,
Independence Square, Philadelphia, PA 19106, USA.

Printed in Great Britain by Galliard (Printers) Ltd, Great Yarmouth, Norfolk

Preface

In 1952 a young mathematics graduate entered the Metallurgy Department of Sheffield University to start his PhD. For his troubles he was assigned an office above the foundry, which was thought to be an ideal place to bring a theorist down to earth. So began Ron Bullough's career in the theory of materials. This volume records the proceedings of a two day Symposium held in honour of the 60th birthday of Ron Bullough, and celebrates contributions to a subject which developed over the four decades of his life. It is helpful to outline Ron Bullough's career and put into context the topic of this meeting.

At Sheffield Ron Bullough's supervisor was Bruce Bilby, now an Emeritus Professor there. It was a stimulating environment with Jock Eshelby in the Department together with many distinguished pioneers in the theory of metals as co-workers and contemporaries. These first years were spent on the theory of continuous distributions of dislocations and its application to interfaces - a theme which recurs several times in later work. The investigations were applied to martensitic transformations and deformation twinning, and have been seminal in guiding subsequent understanding of mismatches at interfaces and surfaces.

In 1955 Ron Bullough moved to the AEI Fundamental Research Laboratory where he was involved in a wide range of studies, including the properties of dislocations in semiconductors, the theory of strain ageing and solute hardening and the modelling of imaging in electron microscopes. It was during this period that Ron Bullough gained the rare reputation as a theorist for seeking out and interacting with experimentalists. Much of his work was in response to experimental observations and was directly useful to AEI projects.

In 1963 Ron Bullough joined Theoretical Physics Division at Harwell Laboratory, the main research establishment of the UK Atomic Energy Authority (AEA). He became Leader of the Radiation Damage and Theoretical Metallurgy Group, a post which he held until 1985. During that period he contributed to the development of the fundamental tools that are used in modelling the behaviour of materials - lattice statics and molecular dynamics for atomistic simulations, Green's functions for fracture studies and the chemical rate theory of point defect kinetics for describing microstructural evolution. These techniques were applied to problems at the heart of the AEA's nuclear programmes and diversification work on industrial problems. The most important of these was probably the understanding of the swelling of structural materials exposed to fast neutron irradiation, which was stimulated by observations in the Dounreay Fast Reactor and threatened to undermine the development of commercial fast reactors. This work led to spin-off developments in the understanding of irradiation creep, irradiation growth of anisotropic materials, nuclear fuel swelling, and helium embrittlement of irradiated austenitic alloys, as well as in the prediction of the behaviour of materials in future fusion reactors. Other problems tackled were less specifically nuclear in character, e.g. the plasticity and fracture of bcc alloys, the growth of epitaxial layers, and the effect of stress driven solute segregation on grain boundary fracture. Many younger theorists and experimentalists benefited from their collaboration with Ron Bullough, and the number of published papers to which his influence can be traced must run into thousands. The development of interatomic potentials, the theory of segregation in alloys and the modelling of nuclear fuel behaviour for normal and accident conditions were examples of activities that flourished in his Group. More surprising was the development and application of economic modelling of energy systems in collaboration with the IEA and the UK Department of Energy.

In 1985 Ron Bullough moved to the position of Head of Materials Development Division (until 1988) and subsequently became AEA Chief Scientist and Director of Corporate Research, which inevitably meant a heavier administrative load. Despite this, he still contributes scientifically to specific areas, two of which are represented in these proceedings. However, his broad knowledge is now being mainly used to shape the research policy of the AEA, and his commitment to ensuring that research is relevant to industrial problems is perhaps his greatest contribution yet.

The concept of the Symposium and of these proceedings was to give an overview of the current development and application of materials modelling in the four major areas which Ron Bullough is associated with: Theory and Observation of Crystal Defects; Radiation Damage in Materials; Modelling Fracture; and Sensors and Electronic Behaviour. All the papers presented to the Symposium were invited from leading exponents in their fields. The speakers were requested to provide concise papers that would be acceptable to a wide audience, and bring out the benefits of modelling to technology. In the event, the papers delivered were of a very high standard. The presentations at the Symposium itself were enthralling; and, regrettably, capturing the humour and spontaneity of the gathering is not possible in these proceedings.

We are fortunate to have an introductory paper by Sir Alan Cottrell which reviews theoretical models of materials, concentrating on an atomistic view of modelling. This was originally given as the concluding paper of the Symposium, but makes a worthwhile introduction to these proceedings. One contribution to the Symposium that is missing here is the very amusing and perceptive appreciation of Ron Bullough's career given by Brian Eyre, Chief Executive of the AEA. Brian Eyre was for many years a collaborator with Ron Bullough, providing the experimental input that stimulated numerous joint investigations. We are very grateful to Brian Eyre and the AEA for support in making the Symposium possible.

In the four areas covered in these proceedings, the papers are ordered, as they were in the Symposium, going from the most basic to the most applied. This reflects the title "Materials Modelling: from Theory to Technology," and emphasises the dynamic nature of the subject which ever requires basic thinking to provide the tools and the stimulus for future applied work. One notable experimentalist, on hearing the Symposium title for the first time, suggested the alternative "Materials Modelling: from Fantasy to Reality?" This may seem cynical, but modelling has had an uphill struggle to become accepted as useful. Looking back, we have come a long way and we think the question mark can be safely removed.

A number of Ron Bullough's key collaborators have contributed to these proceedings, including Ted Smith, a contemporary postgraduate student at Sheffield, Ron Newman who was at AEI and Alan Brailsford who was at Harwell during the most productive years on radiation damage theory. There are also papers from those who started from more junior positions or fellowships in Ron Bullough's Group at Harwell, e.g. Mike Finnis, Eduardo Savino and Sue Murphy: indeed, four of the Editors spent periods of their career under Ron Bullough's guidance. Ron Bullough's own son Tim contributed a paper too, hinting that perhaps a dynasty is in the making! The fact that these proceedings represent a concise and encompassing statement of materials modelling is a clear indication of Ron Bullough's profound influence.

Harwell Laboratory   COLIN ENGLISH
November 1991   JUAN MATTHEWS
                                         HERMANN RAUH
                                         MARSHALL STONEHAM
                                         ROGER THETFORD

Contents

Introduction

Theoretical models in materials science

Sir Alan Cottrell FRS FEng

Department of Materials Science and Metallurgy, University of Cambridge

ABSTRACT: The development of theoretical models in materials science is briefly outlined. For the central problems of the subject, such as those on which Ron Bullough has worked so effectively, where the overall behaviour is determined by the presence, movements and interactions of various crystal defects, atomistic models provide the only way forward, both to make practical progress in technological problems and to gain physical understanding of the fundamental underlying processes. The atomistic models should make maximum use of phenomenlogical input data, but be consistent with quantum mechanical principles.

1. THE UNLIMITED CHALLENGE OF MATERIALS SCIENCE

The modelling of materials from theory to technology is an excellent theme in this celebration of Bullough's contributions to materials science. Someone once said of Rutherford that he was always on the crest of a wave. I think it is much the same with Bullough, who has so successfully managed to be in the right place at the right time. The right place was Harwell, where the emerging technology of civil nuclear power was exposing deep scientific problems of materials in huge numbers. The right time was the 1960s to 80s. Much earlier than this, the basic theory of real materials was too undeveloped to tackle these problems powerfully. Much later , the withering hand of anti-scientific privatisation economics will have blighted this whole area of endeavour. But Bullough got the timing just right; and so contributes splendidly to the development of materials science, both pure and applied, in the nuclear field.

While nuclear power provided the first big opportunity for materials science to show what it could do, society now sees similar opportunities in many directions; in aerospace, electronics, structural engineering, medicine , to name only a few. With most of the basic engineering principles now well worked out and with modern computers solving the governing equations, the main opportunity for further advance has come to depend on the development of improved materials. Some countries, notably the USA and Japan, have realised this. They have seen that the main science for new engineering technology in the 21st century will be materials science; and have set up vast new research and development facilities to get themselves in a strong position for this.

The structures and properties of materials resemble those of biology in their endless variety and complexity. It follows that the opportunities for research in materials science and technology are virtually unlimited, in the sense that the typical materials system contains some 10^{24} atoms and some of its most important properties depend critically on the type and locations of single atoms. The number of alternative structures and variation of properties is then almost endless.   In this situation the limits of research activity are necessarily set by the priorities of society; whether it seeks long-term benefit or short-term financial gain; what kind of a future it wants; what can be afforded; and so on.

The sheer variety of opportunity which stems from the 10^{24} atoms also brings a problem. it would be both impossible and ludicrous to aim at a theory which took account of every one of these atoms. Approximations must be made in which enormous groups of the atoms are represented by some simple collective concept. In other words, the theory has to work with simplified models of real materials.   This modelling also provides endless opportunities, as more and more refined versions are introduced to extend the theory to more intricate structures and properties of materials.

2. CONTINUUM MODELS

The simplest models are of course those that envisage a continuous medium. For solids the classical example is the elastic continuum.   It had two magnificent progenitors: Hooke's law, which emphasised the linearity of infinitesimal deformation, and so opened the door to linear mathematics; secondly, the infinitesimal calculus and its classical field theory, which enables the corresponding theory of elasticity to solve innumerable problems of stress and strain distribution, even as far as Inglis cracks and Volterra dislocations.   Classical continuum theory is necessarily a phenomenlogical theory.   For example, the elastic constants are not deduced but inserted into the theory as measured experimental parameters.

The success of linear elasticity - and the correspondingly successful theory of viscous fluids - has encouraged people to construct other continuum models. In some cases these have assumed quite different phenomenlogical properties, as in the example of the ideal rigid-plastic solid, which has been so useful for metal working technology. In other cases, extra properties have been added to existing ones, as in the theories of viscoelasticity.

There is however a conceptual danger, often overlooked, in investing a continuum with multiple properties. These properties might be mutually incompatible. Laplace showed in the 18th century that surface energy and bulk energy cannot coexist in a true continuum. One or other must be infinite or zero. Similarly, the coexistence of Hookeian elasticity and Newtonian viscosity necessarily implies also the existence of a finite process lifetime, i.e. a non-continuum feature of the material. it follows that viscoelasticity is an impossible property in a true continuum.

Since we all use continuum theories to describe materials which we know to be atomic, the recognition of such incompatibilities would be mere pedantry, were it not for two things. First, the ratios of such properties give one of the first indications of the breakthrough of atomic properties into the macroscopic world of bulk materials.   Second, the inevitable approximations of atomistic models make these seem crude, compared with the formal elegance of continuum theories. This has led some continuum purists to believe that their approach is superior to the atomistic one. Let them beware. Their apparent rigour often rests insecurely on incompatible assumptions about the basic properties of their continua.

3. ATOMISTIC MODELS

The great central ground in materials science is of course held by the various atomistic models. Although the atomistic theory of solids began in the 17th century with Robert Hooke, it did not take off until X-rays had demonstrated the crystalline state. It began seriously in the 1930s with the dislocation model for crystal plasticity and the point defect models for solid-state diffusion and internal chemical reaction. Although atomistic theory has remained partly phenomenlogical, for example taking in the measured energies of point defects in its calculations of the processes of radiation damage, two features have enabled it to take giant strides into our understanding of materials. The first is that the lattice distortion produced by extended defects, such as dislocations and cracks, is long-range and so can be described accurately by linear elasticity. Second, crystallographic constraints largely dictate the forms which lattice defects can take; for a regular structure can go wrong only in regular ways.

As a result, atomistic theories have now been able to achieve incredibly detailed and accurate representations of complex structures and properties. A good example is Bullough's own work (1985) on radiation damage, especially void swelling. In its virtuosity it is grand opera, with a huge cast - dislocations, vacancies, interstitials, grain boundaries, foreign atoms, platelets, tetrahedra, and still others - all on the stage together, all doing their own thing simultaneously and interacting vigorously in every possible way. It has been a great feat, requiring something like the overall grasp of a Verdi, to command all this and bring it to order.

While elasticity and crystallography are extremely useful in atomistic theories, there have always been some problems that required a deeper foundation. Vitek (1985) demonstrated this when he showed how various peculiar features of the plastic deformation of body-centred-cubic metals depend on the atomic structure in the cores of screw dislocations. The need was for a better representation of the forces between atoms; for a theory of interatomic potentials in the solid state. When these problems first arose, for example in the estimation of energies of point defects, grain boundaries, and surfaces, quantum mechanics was not yet ready to deliver such interactions in a simple and usable form for complex materials problems. We had to make do with simple assumptions about bonds between pairs of atoms and take the total energy as a straightforward sum of independent bond energies. Such theories of pair-potentials successfully gave rough indications but no-one was very happy about them.

It thus came as no surprise when Finnis and Sinclair (1985), from Bullough's Theoretical Physics Division at Harwell, pointed out some fundamental faults of the pair-potential model, even when this had be(an realistically improved, for metals, by including a major volume-dependent term to represent the energy of the free electrons. The important positive contribution of their work was to replace the over-simple pair-potential by a many-body one which, although of impeccable quantum-mechanical parentage, is almost as easy to use as before. You simply count the number of bonds an atom makes with its near neighbours and then take the square root as a measure of the cohesive energy.

4. QUANTUM MECHANICAL MODELS

This brings us to the ultimate: quantum mechanical calculations, the possibility of solutions direct from Schrodinger's equation or its more practical representative, the density functional equation. In principle this should mean emancipation at last from simplified models. Simply feed in the atomic numbers of the participating atoms and get the answers from the equation. In practice it is now possible to go a long way towards realising this 'ab initio' approach with the help of big computers. There have for example been some spectacular calculations of the cohesive properties of transition metals (Moruzzi et al, 1978).

But there is an intriguing aspect to all this. The computer is fed some numbers and then delivers some other numbers, such as the cohesive energy of iron. What have we learnt from this? That Schrodinger's equation or its representative is still performing well. But we might have expected this, anyway. We have learnt nothing about iron, since we already knew its cohesive energy from measurement, beforehand. The realisation of this must surely change our attitude to models. Whereas before, we had to use them because there was nothing better, now that we can in principle dispense with them altogether we see that they are, after all, what we really want. We want a physical picture of what is going on in materials, an understanding of the main effects determining structures and behaviour, backed up by numerical agreement to confirm the validity of our picture. This is really a philosophical point: physical understanding comes only from simplified models and always will do so.

The impact of the quantum theory of solids on materials science is extraordinary. All the great principles of the theory were established before 1940 - Bloch waves, Fermi surfaces, Brillouin zones, etc. - yet they have had little effect on materials science. By contrast the post-war developments, which have been more concerned with detailed techniques of calculations than general principles, are having a major impact.

Before going on to this, two exceptions to what I have just said should be noted. First, the pre-war Brillouin zone theory of alloy phases. This became holy writ for physical metallurgists until Pippard in 1957 showed experimentally that the Fermi surface already touched the Brillouin zone boundary in pure copper, whereas it was not expected to do so until the limit of primary solid solution was reached. A further blow came from the analysis of Heine and Weaire (1970) who showed that even when the Fermi surface touched a zone boundary no sharp variation of energy with electron concentration was to be expected.

The second exception was the post-war Mott-Hubbard theory which changed our ideas of what makes some materials metals and others insulators. Even more importantly, this theory weaned people away from over-dependence on Brillouin zone concepts and caused them to emphasise the distribution of the electrons in real space, among the atoms, rather than in reciprocal space. This change in emphasis underlies much of the post-war development and is partly responsible for the recent increase in the utility of the quantum theory of solids.

The three main developments in techniques which have been useful to materials science are: the simplification of the tight-binding theory by the moment distribution method, which is responsible amongst other things for the convenient square-root expression of the many-body potential; (2) the pseudopotential theory, which in the hands of Heine and Weaire (1970) put new life into Brillouin zone explanations of various structures and properties of non-transition metals; and (3), above all, the density functional theory which has freed the quantum mechanical approach from its previous limitation to perfect crystal structures and so enables it to begin to deal with the problems of real materials.

Density functional theory is a grand one which can be used at various levels, from the ab initio at one extreme to simplified models at the other. One of the most successful of its models is the effective medium version of the embedded atom theory (Nørskov 1982a). Here the model is a spatially uniform electron gas, of density equal to the local value in the site in which the atom of interest is embedded. Most of the energy of interaction of this atom with the surroundings then comes from its interaction with this effective medium, the values of which depend only on the electron density and are thus 'universal' quantities for each species of embedded atom, which have in many cases been calculated and published. By this means it has been possible to show, for example, that a hydrogen atom is not only attracted into a vacancy, in iron, but that, in the vacancy, it sits at 0. 5A to one side of the centre, because that is where the electron density is at the optimum for hydrogen (Nørskov 1982b). This is a virtuoso result in theoretical materials science!

5. ATOMIC SIZES IN METALS AND ALLOYS

One concept which brings the atomistic and quantum mechanical models together is that of atomic size. It led of course to Hume-Rothery's brilliant first generalisation, the limit on solid solubility when the atomic radii differ by more than 15%, which has been explained from elementary elasticity and thermodynamics. But more detailed work showed that the concept was not a simple one. It depended for example on both ionic and atomic radii, from which Hume-Rothery and Raynor's useful classification into 'full' and 'open' metals stemmed. In particular, Raynor (1949) showed that the changes in lattice parameter, when higher valency elements are dissolved in copper and silver, are contributed by two factors: intrinsic atomic (or ionic) size; and valency. The effective medium theory points to something similar. For each element there is an ideal electron density at which its embedding energy is most favourable; and it will distort its surroundings in order to approach this ideal, so much so in extreme cases as to produce a complex structure such as those of icosahedral symmetry (Redfield and Zangwill, 1988).

The distinction between atomic size as indicated by ionic radius and that governed by the conduction electron density has come into prominence in recent quantum-mechanical models. Thus Hafner and Heine (1983) showed that the pseudopotential core radius (approximately equal to the ionic radius) , atomic radius, and valency, jointly determine whether a crystalline element is close-packed and metallic, or more open-packed and covalent. Very recently, Pettifor (1989, 1991) has developed a new many-body potential theory which generalises the previously mentioned square-root term, of the tight binding scheme, to take account of three-body and higher interactions to predict trends in structural behaviour through the periodic table.

6.   ROLES OF QUANTUM MECHANICAL AND ATOMISTIC MODELS IN MATERIALS SCIENCE

Finally, a word on the respective roles of quantum mechanical and atomistic models in materials science. Obviously, only quantum mechanics can explain what makes iron different from copper, or why the elastic constants of aluminium have their particular values. But in the more typical materials science problems, where the interactions of lattice defects are usually involved, as for example in Bullough's radiation damage studies, then atomistic modelling is the only practical way forward. Occasionally, people who have been frustrated by the sheer difficulty of some of the materials science problems, such as what goes on at the tip of a crack in a metal such as iron, have turned in desperation away from atomistic towards quantum mechanical models. think this may sometimes be a mistake. The difficulty in such problems usually lies, not in the fundamental atomic interactions, but in t.he sheer geometric and atomistic complexity of the situation. Such difficulty has to be overcome on its own terms by improving the atomistic modelling, as Bullough (1985) showed in his void growth work, and Hirsch (1991) has recently shown in his explanation of the anomalous temperature dependence of the yield strength in some intermetallic compounds.

As far as possible in these atomistic models, the input data should be phenomenological, from measured values of energies, defect mobilities etc. Not to take advantage of such data is simply to make the problem unnecessarily hard. Occasionally however, it is not possible because the data are unavailable or may even be experimentally inaccessible. It is then necessary to turn to quantum mechanics to fill the gap. I recently came across an example of this in a study of the effect of interstitial impurity atoms on intergranular cohesion in transition metals (Cottrell 1990). People have tried to tackle this problem from measured values of the chemisorption energy of such impurities on the fracture faces, but these measurements represent a situation in which the atoms have had tune to find equilibrium positions, in some cases by surface reconstruction, whereas in fast intergranular fracture at low temperature there is no opportunity for this, and the exposed interstitial atoms will be left sitting on the surface in unstable positions as the crack passes through. Their state of cohesion during this process is thus experimentally inaccessible by today's techniques, and so has to be estimated quantum mechanically.

To conclude, in the characteristic problems of materials science, use atomistic models as far as possible, provided that these are demonstrably compatible with quantum mechanical principles. But then, that is just what Bullough has always done!

REFERENCES

Bullough R 1985 in Dislocations and Properties of Real Materials,
The Institute of Metals (London) p 283
Cottrell A H 1990 Mater.Sci.Tech. 6 121
Finnis M W and Sinclair J E 1985 Phil.Mag. A50 45
Hafner J and Heine V 1983 J.Phys.F. : Metal Phys. 13 2479
Heine V and Weaire D 1970 Solid State Phys. 24 249
Hirsch P B 1991 Prog.Mater.Sci. in press

Moruzzi V L et al 1978 Calculated Electronic Properties of Metals
Pergamon Press (Oxford)
Nørskov J K 1982a Phys.Rev. B26 2875
Nørskov J K et al 1982b Phys.Rev. Lett. 49 1420
Pettifor D G 1989 Phys.Rev.Lett. 63 2480
Pettifor D G and Aoki M 1991 Phil.Trans.Roy.Soc. A334 439
Raynor G V 1949 Trans.Farad.Soc. 45 698
Redfield A C and Zangwill A 1988 Phil.Mag.Lett. 57 255
Vitek V 1985 in Dislocations and Properties of Real Materials
The Institute of Metals (London) p 30

Part 1
Theory and Observation of
Crystal Defects

The continuized crystal as a model for crystals with dislocations

E Kröner

Universität Stuttgart und Max-Planck-Institut für Metallforschung Stuttgart

ABSTRACT: The concept of continuized crystal is introduced as a model which conserves the main characteristics of a crystal and, at the same time, permits the use of continuous functions. During the continuization process the distance between the (scaled-down) dislocations, when measured in (scaled-down) atomic spacings a, increases. In the limit $a \to 0$, the crystal containing a macroscopically homogeneous distribution of dislocations is force-stress free, but contains localized moment stresses at the positions of the individual dislocations. Thus moment stress is the specific response to dislocations. It can be calculated by use of the theory of Peierls and Nabarro. The Peierls-Nabarro stress as a limit stress as well as the moment stress and potential energy, both taken per length of the Burgers vector, are not changed in the transition, which ensures the physical applicability of the concept of continuized crystal.

1. DISCRETE VS. CONTINUOUS DESCRIPTION OF DISLOCATIONS. CONTINUIZED CRYSTAL.

Ron Bullough whose 60th birthday is celebrated today has given many essential, even some fundamental, contributions to the field of defects in ordered structures, in particular crystals. One of them is the admirable introduction of differential geometry for the representation of dislocations (together with B.A. Bilby and E. Smith 1955), another the application of lattice theory for a similar purpose - the description of crystal defects (together with J.R. Hardy 1966).

Whereas differential geometry deals with continua (in a certain sense, see below), lattice theory emphasizes the discrete aspects of crystalline matter. The question then arises as to which extent can crystalline matter be described by a continuum picture. Whereas, doubtless, there are phenomena, such as short range defect interaction, which require a discrete description, there are long range phenomena whose characteristic lengths are large compared to the atomic spacing. The feeling is that in such situations a continuum description could work quite well.

Continuous and discrete description have a similar mutual relation as do classical and quantum mechanics. To understand this, recall that the crystal occurs in so-called *configurations* specified by the positions of all particles. Let's call them atoms. These particles, however, have momenta. We can then use Hamilton's equations to build up a quantum version of the classical lattice dynamics. Note the spectacular symmetry between the two sets of Hamilton's equations. Now the point: This symmetry is also visible if differential geometry is used to describe the geometric, or kinematic, part of dislocation theory. Since the static response to any distribution of dislocations should be equal to the variational derivative, with respect to this distribution, of structure, i.e. obey the same type of differential-geometrical equations. This has been proved in sufficient generality by Stojanović (1963, linearized theory) and by Kröner (1987, nonlinear theory).

The transition from quantum towards classical mechanics can be managed by means of the limiting process $h \rightarrow 0$, with h the Planck's constant. Similarly we introduce for our purpose a limiting process $a \rightarrow 0$, where a is the lattice spacing. To simplify the story, consider Bravais crystals only. Imagine that a similar process is performed on the static side of the theory. For our present purpose details of this process are not relevant.

The imagined body obtained by the limiting process $a \rightarrow 0$ has been called *continuized crystal* by Kröner (1986). It is not a continuum in the strict sense of continuum mechanics, because the most important characteristics of a crystal are conserved in the limiting process. There is (i) the existence, at every point, of three nonplanar crystallographic atomic rows and (ii) the possibility of atomic step counting along these rows, which means that the continuized crystal has a *metric* structure. (This structure is lost, if vacancies and/or self-interstitials are introduced.)

In the limiting process all atoms are scaled down such that the total mass per unit volume remains unchanged. At the same time the dislocations are scaled down to conserve the relation between Burgers vector b (of length b) and atomic spacing $a(= b)$. The total amount of Burgers vector per unit area must not be changed in this process. It is under these conditions only, that we can hope to obtain a good model for the real physical crystal.

There is a peculiar consequence of these conditions. In Figure 1a,b we see the crystal in two stages of the indicated limiting process. The lattice spacings in 1a are 4 times those in 1b. The distance d between neighboring dislocations in 1a is 2 times that in 1b. This means that (b/d) in 1b is one half times b/d in 1a. After many such steps of the limiting process we arrive at $(b/d)/n$ with n a large number. Note that the content of Burgers vector remains the same in this process. This requirement causes the distance between neighboring dislocations to grow (!) by a factor of 2 in Figure 1, when measured in (scaled) atomic spacings. Of course, the (scaled) dislocations move closer when observed on the macroscale. In the limit $a \rightarrow 0$ the distance between neighboring dislocations becomes an infinite number of atomic spacings. That means that the scaled down dislocations are very discrete on the (scaled) atomic scale, although they form a density in the macroscopic observation.

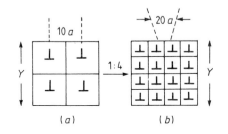

Fig. 1. Two stages in the continuization of a crystal with dislocations.

The conclusion from these findings is that the differential geometry introduced for the description of dislocations by Kondo (1952) and, independently, by Bilby, Bullough and Smith (1955) describes the continuized crystal with dislocations defined above. This crystal is not a continuous piece of matter in the sense of conventional continuum mechanics. With Cottrell (1986) we may say that the transition "$b \rightarrow 0$, which means $a \rightarrow 0$, leads only to a countable infinity of points, not to the higher infinity of the continuum."

2. DISLOCATIONS AND MOMENT STRESSES

In the continuum picture the presence of dislocations causes strains around the line and, as a response to these, stresses as known from conventional elasticity theory. These stresses are defined by the contact forces transmitted through internal area elements. We speak of *force stresses* to distinguish them from the so-called *moment stresses* (transmission of moments) which also play a role in the theory of dislocations. It will be shown here, how this comes along. We partly follow older work of Kröner (1963) and of Hehl and Kröner (1965).

The stress field of a straight edge dislocation of Burgers vector be_1 and line vector te_3 in an infinite medium and in linear approximation is

$$\sigma_{xx}(x,y) = Ay(3x^2 + y^2)/\rho^4, \quad \rho^2 \equiv x^2 + y^2, \quad (1)$$

where (x,y,z) is a cartesian coordinate frame with base vectors e_1, e_2, e_3 and $A \equiv \mu b/2\pi(1-\nu)$, μ,ν the shear modulus and Poisson's ratio respectively. The also present stresses σ_{xy} and σ_{yy} are here of no interest.

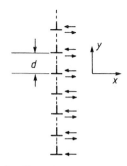

Consider the stress response in the plane $x = 0$. This stress is $\sigma_{xx}(0,y) = A/y$ and has a resulting moment around the z-axis on the plane $x = 0$

$$\int_{-\infty}^{\infty} \int_{-\infty}^{\infty} y\sigma_{xx}(0,y)dydz = A \int_{-\infty}^{\infty} dy \int_{-\infty}^{\infty} dz \quad (2)$$

so that A is the averaged moment per unit area of the plane $x = 0$. The fundamental feature of this result is that a moment appears in the repsonse to the presence of a dislocation.

Fig. 2. Small angle grain boundary and moment stress.

Next consider a whole wall of edge dislocations in the so-called small angle grain boundary array (Figure 2). For symmetry reasons the stress is periodic with period d in y-direction. Summing up the contributions of all dislocations we obtain for the interval $-d < y < d$ the stress

$$
\begin{aligned}
\sigma_{xx}(0,y) &= A\left(\frac{1}{y} + \frac{1}{y+d} + \frac{1}{y+2d} + \ldots - \frac{1}{d-y} - \frac{1}{2d-y} - \ldots\right) \\
&= [\mu/2\pi(1-\nu)](\frac{b}{y} + \frac{1}{(y+d)/b} + \frac{1}{(y+2d)/b} + \ldots \\
&\quad - \frac{1}{(d-y)/b} - \frac{1}{(2d-y)/b} - \ldots).
\end{aligned}
\tag{3}
$$

If the wall is one of the continuized crystal, then $d/b \to \infty$ and $\sigma_{xx}(0,y)$ vanishes for all $y \neq 0$ in the interior of the interval $-d\ldots + d$. Thus for the continuized crystal (3) reduces to

$$
\sigma_{xx}(0,y) = \begin{cases} [\mu/2\pi(1-\nu)]b/y & \text{for} \quad y = 0 \\ 0 & \text{for} \quad y \neq 0, \end{cases}
$$

valid for the mentioned interval: $\sigma_{xx}(0,y)$ vanishes for all $y \neq 0$. If b/d is small, but finite, then $\sigma_{xx}(0,+0)$ and $\sigma_{xx}(0,-0)$ are near infinity and of opposite sign and we expect a finite moment. This moment goes down when b/d goes down.

We have found, that in the case of a single dislocation wall as described the resulting stress along the plane $x = 0$ vanishes except at the positions of the dislocations. Here it forms moments as shown in Figure 2. The resulting moment on a part of the plane $x = 0$ is proportional to the size of this part, which means that a moment stress can be defined. Clearly, there are the same moments, only of opposite sign, when the plane $x = 0$ is considered, whose normal unit vector is opposite of the one considered before. Next we argue that in a macroscopically homogeneous arrangement of dislocations in a continuized crystal we have no stresses except moment stresses at the positions of the dislocations. At this time our argument is convincing only for *random* (macrohomogenous) dislocation distributions. In this case, any plane $x = $ const. will contain a number of dislocations in a small angle boundary array. For an infinite crystal the relative fluctuation of this number will be zero.

Since the stress field of a dislocation has an equal amount of positive and negative parts, it is plausibel that in a really random distribution there will be a lot of cancelling of the stress from the dislocations. In the limit of the continuized crystal we therefore expect zero stress everywhere except at the positions of the dislocations, where we should have a localized moment stress. This means, that at all positions where the dislocation density is non zero, also moment stresses are present. In other words, moment stresses are the specific response to dislocation density. This result is close to a theorem by Nye (1953) according to which a homogeneously distributed dislocation density causes a homogeneous curvature of the crystal lattice structure. It is natural to assume that moment stress should arise as the response to such curvature.

3. MOMENT STRESS AND PEIERLS POTENTIAL

In eq. (3) the stress in the plane through a dislocation wall was given. In the continuized crystal this stress vanishes except at the positions of the dislocations, where it gives rise to a localized moment stress. Obviously, the size of this moment stress cannot be calculated from linearized elasticity theory. A nonlinear theory such as that of Peierls (1940) and Nabarro (1947) is needed.

The notation used in this model is explained in Figure 3, where, for simplicity, a primitive cubic lattice is chosen. Whereas the half-spaces **A** and **B** are treated as linearly elastic, the interaction between the atomic planes A and B, adjacent to the glide plane, is described by the Peierls potential. This potential is periodic in the relative displacement $u^{AB}(x) = u^A(x) - u^B(x)$ of the points in A relative to those in B. The negative derivative of the potential is the shear stress $\sigma_{yx}(x)$ which we assume in the well-known form

$$\sigma_{yx}(u^{AB}) = \frac{\mu}{2\pi} \sin \frac{2\pi u^{AB}(x)}{b}. \tag{4}$$

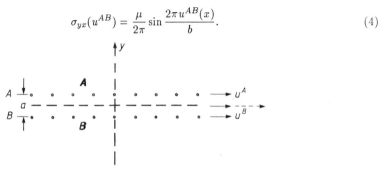

Fig. 3. Nonlinear dislocation model of Peierls and Nabarro.

Multiplied by the area element $dxdz$ of A, (4) is the force acting on the area element of B, and minus the force on the corresponding element of A, whose normal vector points downwards. The two forces together build up a moment $b\sigma_{yx}dxdz$ around the z-axis.

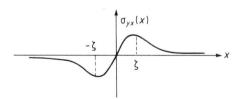

Fig. 4. Shear stress in the glide plane according to Peierls and Nabarro.

Peierls solution for the edge dislocation with half-plane inserted above A is

$$u^{AB}(x) = \frac{b}{\pi}\tan^{-1}\frac{x}{\zeta} \qquad (5)$$

where $\zeta = b/2(1-\nu)$ is called the width of the (edge) dislocation. As follows from (4) and (5), $x = \zeta$ is the point where the curve $\sigma_{yx}(x)$ has its maximum $\mu/2\pi$, whereas the minimum $-\mu/2\pi$ lies at $x = -\zeta$ (Figure 4). The resulting moment per unit length of the dislocation, taken between $x = 0$ and $x = x_0$ is

$$m_z = \int_0^{x_0} b\sigma_{yx}dx = \frac{\mu b}{\pi}\int_0^{x_0}\sin(2\tan^{-1}\frac{x}{\zeta})dx = \frac{\mu b\zeta}{\pi}\ln(1+\frac{x_0^2}{\zeta^2}). \qquad (6)$$

The integral in (6) diverges for $x_0 \to \infty$, since the shear stress (4) with (5) decays like $1/x$ for large x. However, in the continuized crystal the far-fields of the dislocations cancel mutually, so that the integration may be extended only finitely, e.g. to $x_0 = 2\zeta$. For a Poisson number $1/3$ we then obtain $m_z = \alpha\mu b^2$ with $\alpha \approx 0,17$. To obtain the correct α of our model, we should make the Peierls calculation for *arrays* of dislocations.

$\Delta m_z = \tau_{zx}Y$ defines the moment stress τ_{zx} in our special situation. Δm_z is the resulting moment per unit length of the dislocations in Figures 1a and b. It is *not* scale invariant, but τ_{zx}/b is (check!) and this suffices. The analogous result applies to the potential energy E which also has a b^2-dependence. Thus E/b is scale invariant. Together with that of the Peierls-Nabarro stress these invariances testify to the physical reality of the model of continuized crystal.

4. CONCLUSION

The smallness of the lattice parameter (a) of Bravais crystals, those to which we have restricted ourselves for simplicity, suggests to make the transition $a \to 0$ such that the mass and defect content per unit volume and unit surface remains unchanged. The result of this scaling process is the continuized crystal, not a continuum in the ordinary sense. Nabarro's (1947) expression for the Peierls stress of a primitive cubic lattice shows that this stress remains invariant in the transition as emphasized earlier by Cottrell (1986). The main results of this paper are (i) new arguments for the existence of moment stresses which represent the specific response to dislocation distributions and (ii) the proof that the resulting moment on a unit internal surface as well as the potential energy, both taken per length of Burgers vector, are invariant under the scaling. These results confirm the physical applicability of the concept of continuized crystal. Note that the constitutive law relating moment stress to dislocations, $m_z = \alpha \mu b^2$ in our example, depends above all on the Peierls potential. This means, it cannot be calculated from elastic moduli only, not even from those of the non-linear theory.

The considerations of this work might appear somewhat academic. In fact, it is not the aim of our effort to develop a method for easy calculation of the behaviour of crystalline solids. The motivation is rather that the similarity between dislocations in crystals on the one' hand and torsion in differential geometry on the other hand is so striking that there must be some physical reality behind it. If, however, differential geometry with torsion describes physical reality, then the question arises as to what is the material manifold which obeys the equations of differential geometry (as we have argued, on both the kinematic and the static side). It was found that exactly this is the continuized crystal.

5. REFERENCES

Bilby B A, Bullough R and Smith E 1955   Proc. Roy. Soc. (London) **A231** 263

Cottrell A H 1986   S. Afr. J. Phys. **9** 44

Hardy J R and Bullough R 1967   Phil. Mag. **15** 237

Hehl F and Kröner E 1965   Z Naturforschung **20a** 336

Kondo K 1952   Proc. 2nd Japan Nat. Cong. Appl. Mechanics pp. 41-7

Kröner E 1963   Int. J. Engng. Sci. **1** 261

Kröner E 1986   Z. Ang. Math. Mech. **66** T 284

Kröner E 1987   Phys. Stat. Sol (b) **144** 39

Nabarro F R N 1947   Proc. Phys. Soc. (London) **59** 256

Nye J F 1953   Acta Met. **1** 153

Peierls R E 1940   Proc. Phys. Soc. (London) **54** 34

Stojanović R 1963   Int. J. Engng. Sci. **1** 323

Deformation twinning in 4-dimensional lattices

A G Crocker

Department of Physics, University of Surrey, Guildford GU2 5XH

ABSTRACT: Twinning is an important deformation mechanism of crystalline materials. The classical crystallographic features of twinning, corresponding to Type I, Type II and Compound modes, are first summarised. A generalised theory, which results in seven groups of twins, is then presented. This theory is formulated using the notation of tensor calculus. When the superscripts and subscripts of this notation range from 1 to 3 both classical and non-classical twinning modes in 3-dimensional crystals are predicted. However, when the range 1 to 4 is adopted, results for 4-dimensional lattices are obtained. Examples of twinning modes in 4-d cubic lattices are presented and discussed.

1. INTRODUCTION

Ron Bullough and I shared the same PhD supervisor, Bruce Bilby. However, when I joined Bruce at Sheffield in 1956, Ron had already left and had been working at AEI for a year. The Bullough and Bilby (1956) theory of martensite crystallography had just been published and Bruce asked me to apply this analysis to a range of transformations. I had already been introduced to lattice geometry, dislocation theory and deformation twinning while doing an undergraduate project with Maurice Jaswon at Imperial College and as many martensite plates are twinned I was also able to extend this interest. I did not however get deeply involved in the theory of continuous distributions of dislocations which Bruce, Ron and Ted Smith had developed (Bilby et al 1955) and at the time was being pursued by my fellow research student Les Gardner. At AEI, Ron had taken an interest in twinning modes in semiconductor materials (Bullough 1957) so there were several reasons for us to interact.

On leaving Sheffield I joined Battersea College of Technology which in 1966 became the University of Surrey, and in 1962 took on two of my own research students David Bacon and Mike Bevis. Dave investigated the elastic energies of various dislocation configurations, particularly loops (Bacon and Crocker 1966), and Mike tackled problems in lattice geometry, concentrating on generalising theories of deformation twinning (Bevis and Crocker 1968). During the period of their PhD projects I had a sabbatical year in the States and was fortunate to be able to arrange for Ron Bullough to help with their supervision. As a result of this Ron became a Visiting Professor at Surrey and interacted with us closely for several years. In particular he helped us to develop research projects on computer simulation of crystal defects including of course twin boundaries. It is however the lattice geometry of deformation twinning which is the theme of the present paper.

2. DEFORMATION TWINNING MODES

Following Bilby and Crocker (1965), a deformation twinning shear is defined to be a homogeneous shear which restores the lattice in a new orientation. A schematic example is shown in Figure 1, where the central section of a lattice is sheared homogeneously on a plane represented by K_1 in a direction η_1 contained in K_1. For a particular value of the shear strain g, the lattice is reproduced in a new orientation. Figure 1 also shows that a second homogeneous shear on the plane K_2 in the direction η_2, but with the same shear strain g, also restores this lattice. It is convenient (Cahn 1965) to represent these two conjugate twinning modes by the elements K_1 K_2 η_1 η_2; g or equivalently by $\mathbf{m_1}$ $\mathbf{m_2}$ $\mathbf{l_1}$ $\mathbf{l_2}$; g, where $\mathbf{m_1}$ and $\mathbf{m_2}$ are unit vectors perpendicular to K_1 and K_2 and where $\mathbf{l_1}$ and $\mathbf{l_2}$ are unit vectors parallel to η_1 and η_2. These elements are not of course independent as any mode or pair of modes is defined by only four degrees of freedom. For example, $\mathbf{m_1}$ and $\mathbf{m_2}$, each including two degrees of freedom, completely define a mode as does $\mathbf{m_1}$, $\mathbf{l_1}$ and g where, as $\mathbf{l_1}$ lies in $\mathbf{m_1}$, $\mathbf{m_1}.\mathbf{l_1} = 0$.

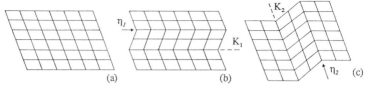

Figure 1. Homogeneous shear of a lattice (a), on the plane K_1 in the direction η_1 to produce a twin (b) and on the plane K_2 in the direction η_2 to produce the conjugate twin (c).

Normally K_1, K_2, η_1 and η_2 are represented by Miller indices and, noting that the orientation relationships of Figure 1 can be described as either reflection in K_1 or rotation of π about η_1, and similarly for K_2 and η_2, the three classical types of twinning mode arise. These are Type I in which K_1 and η_2 have rational Miller indices, but K_2 and η_1 are irrational, Type II in which K_2 and η_1 are rational but K_1 and η_2 are not, and finally compound which is a combination of Types I and II in which all four elements are rational (Cahn 1954). Here the irrational elements are functions of the lattice parameters and always degenerate to become rational in cubic lattices and usually do so in other high symmetry lattices. Thus for example the twinning mode in body centred cubic metals is

$$\{112\} \ \{\bar{1}\bar{1}2\} \ <\bar{1}\bar{1}1> \ <111> \ ; \ 0.707.$$

However in low symmetry lattices irrational elements do arise, the observed mode in crystalline mercury, which has a rhombohedral crystal structure, being

$$'\{\bar{1}35\}' \ \{\bar{1}11\} \ <\bar{1}21> \ '<0\bar{1}1>'; \ 0.633$$

where the apostrophes indicate rational approximations to irrational elements (Guyoncourt and Crocker 1968). It is also possible for twinning modes with three rational elements to arise in low symmetry crystals (Crocker 1965) but these have not been observed.

3. PREDICTION OF TWINNING MODES

A generalised theory of the crystallography of deformation twinning based on the definition that a twinning shear restores the lattice in a new orientation has been developed by Bevis and Crocker (1968). In this theory the shear is represented by the matrix S which, as shown in Figure 2, relates a vector $\mathbf{x}$ of the parent lattice to the sheared vector $\mathbf{y}$ of the twinned lattice by means of the equation

$$\mathbf{y} = \mathbf{Sx} \tag{1}$$

However, again as shown in Figure 2, $\mathbf{y}$ is also related to a vector $\mathbf{z}$ of the parent lattice by means of a rotation $\mathbf{R}$ so that $\qquad \mathbf{y} = \mathbf{Rz}.$ $\qquad\qquad\qquad$ (2)

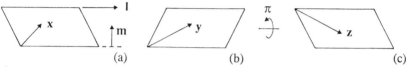

Figure 2. Homogeneous shear of a unit cell (a) to produce a twin cell (b) which can also be generated (c) by a rotation of π of the original cell.

Finally, as they are vectors of the same lattice, $\mathbf{z}$ and $\mathbf{x}$ must be related by the equation

$$\mathbf{z} = \mathbf{Ux}, \qquad (3)$$

where, as no volume change has occurred, the correspondence matrix $\mathbf{U}$ is unimodular, meaning that its determinant is unity. Also as $\mathbf{x}$ and $\mathbf{z}$ are lattice vectors, the elements of $\mathbf{U}$ are integers. Combining equations (1) - (3) we obtain

$$\mathbf{S} = \mathbf{RU} \qquad (4)$$

and multiplying each side of (4) by its own transpose and using the identity $\mathbf{R}^T\mathbf{R} = \mathbf{I}$, the unit matrix,

$$\mathbf{S}^T\mathbf{S} = \mathbf{U}^T\mathbf{U}. \qquad (5)$$

As $\mathbf{S}^T\mathbf{S}$ and $\mathbf{U}^T\mathbf{U}$ are symmetric (5) provides six scalar equations for the four unknowns of the shear matrix $\mathbf{S}$. These can be solved subject to the constraints that $\mathbf{U}$ is unimodular and that the sums of the squares of the elements of $\mathbf{U}^T\mathbf{U}$ and $\mathbf{U}^{-1T}\mathbf{U}^{-1}$ are equal. This second condition comes from the detailed analysis of Bevis and Crocker (1968). Equation (5) then provides two solutions for the twinning plane, two solutions for the corresponding twinning direction and a unique solution for the shear strain, giving the elements K_1 K_2 η_1 η_2; g of a full twinning mode.

Before examining the character of these solutions it is necessary to note that, given a unimodular matrix $\mathbf{U}$, further unimodular matrices may be obtained by changing the signs of rows, changing the signs of columns, permuting rows and permuting columns. For a 3 x 3 matrix this gives rise in general to 576 distinct unimodular matrices and these will be termed equivalent. All of these matrices will predict different twinning modes in triclinic lattices. However the number of variants reduces as the symmetry increases and for cubic lattices all 576 variants predict crystallographically equivalent twinning modes. It is also important to note that if $\mathbf{U}$ is unimodular, so are $\mathbf{U}^{-1}$, $\mathbf{U}^T$ and $\mathbf{U}^{-1T}$ and, subject to the second condition on $\mathbf{U}$ being satisfied, these will predict distinct twinning modes.

It is readily shown (Bevis and Crocker 1968) that all classical twinning modes (Cahn 1954) satisfy the condition $\mathbf{U} = \mathbf{U}^{-1}$. In addition some classical modes satisfy the further conditions $\mathbf{U} \equiv \mathbf{U}^{-1T}$ and $\mathbf{U} \equiv \mathbf{U}^T$. Here the equivalence sign implies equality apart from changes of sign and positions of rows and columns as discussed above. This additional condition results in the equivalence of some of the Miller indices of the twinning elements as illustrated by the observed mode for body centred cubic metals given in Section 2 (Bevis and Crocker 1968, 1969). The striking result of the theory however is that in addition to these two types of twinning mode, a further five groups arise. These do not satisfy the classical orientation relationships and in general have all four twinning elements, K_1, K_2, η_1 and η_2, irrational. The conditions on $\mathbf{U}$, $\mathbf{U}^{-1}$, $\mathbf{U}^{-1T}$ and $\mathbf{U}^T$ which give rise to the total of seven twinning groups are as follows:

1. None;   2. $\mathbf{U} = \mathbf{U}^{-1}$;   3. $\mathbf{U} \equiv \mathbf{U}^{-1}$;   4. $\mathbf{U} \equiv \mathbf{U}^{-1T}$;   5. $\mathbf{U} \equiv \mathbf{U}^T$;
6. $\mathbf{U} \equiv \mathbf{U}^{-1} \equiv \mathbf{U}^{-1T} \equiv \mathbf{U}^T$;   7. $\mathbf{U} = \mathbf{U}^{-1} \equiv \mathbf{U}^{-1T} \equiv \mathbf{U}^T$.

Here cases 2 and 7 are the classical modes. Geometrically possible examples of all seven groups have been found and some of the non-classical cases may have been observed (Bevis and Crocker 1969).

In order to predict which of these modes are likely to occur in practice, physical criteria have to be applied. The first of these is that twinning modes with small shear strains should be preferred. Others are concerned with the fraction of atoms which are sheared directly to twin sites and the shuffling mechanisms which are necessary for the remainder. Further considerations are the energies of the interfaces created and the nature of any steps or twinning dislocations in these interfaces.

4. TWINNING MODES IN 4-DIMENSIONS

The outline of the Bevis and Crocker theory of the crystallography of deformation twinning given in Section 3 is presented using standard matrix notation. The theory itself was formulated using tensor notation in which, for example, the homogeneous shear S is given by $S^i_j = \delta^i_j + g l^i m_j$, where $i, j = 1,2,3$. Note that both superscripts and subscripts are needed as l is a vector of the direct lattice with basis c_i whereas m is a vector of the reciprocal lattice with basis c^i. Similarly the rotation matrix R and the correspondence matrix U are represented by R^i_j and U^i_j and equation (5) which gives the solutions of the theory becomes

$$c_{ij}S^i_k S^j_l = c_{ij}U^i_k U^j_l. \tag{6}$$

Here c_{ij} is the metric of the direct lattice and is given by $c_{ij} = c_i.c_j$ and the summation convention of the tensor calculus has been adopted. The solutions for g, m_i and l^i are then given by

$$g^2 = c_{ij}c^{pq}U^i_p U^j_q - \delta^i_i \tag{7}$$

$$(c_{\alpha\alpha} - c_{ij}U^i_\alpha U^j_\alpha)m^2_\beta - 2(c_{\alpha\beta} - c_{ij}U^i_\alpha U^j_\beta)m_\alpha m_\beta + (c_{\beta\beta} - c_{ij}U^i_\beta U^j_\beta)m^2_\alpha = 0 \tag{8}$$

$$2gm_\alpha c_{\alpha i}l^i = c_{ij}U^i_\alpha U^j_\alpha - c_{\alpha\alpha} - g^2 m^2_\alpha, \tag{9}$$

where $c^{ij} = c^i.c^j$ is the metric of the reciprocal lattice and the summation convention is suspended for Greek letters.

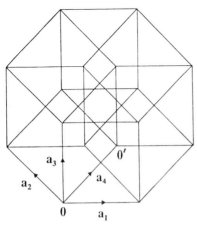

Figure 3.
A 4-dimensional unit cube.

In order to apply the theory to investigate deformation twinning modes in 4-d lattices the superscripts and subscripts of equations (6) - (9) are simply allowed to take on the values 1 to 4 rather than 1 to 3. Here this will be illustrated by considering the simple cubic 4-d lattice. This has four mutually perpendicular basic vectors of the form $a_i = <0001>$ and the unit cell can be represented conveniently as shown in Figure 3. This consists of a 3-d cube with origin at 0 at the bottom left with its eight corners joined to an identical 3-d cube with origin at 0′ at the top right. The vectors a_1, a_2 and a_3 define the cell edges of the first cube and a_4 is the translation vector to the second cube. The resulting 4-d hypercube has 16 corners, 32 edges, 24 faces and 8 solids. The corners, edges and faces correspond to familiar features in 3-d unit cells and the solids are the eight 3-d cubes contained within the 4-d hypercube. Consequently, three additional Bravais lattices can be generated from 4-d simple cubic as the twenty-four faces, eight solids and one hypercube can be centred.

As an example of discovering a twinning mode in a 4-d cubic lattice consider the unimodular correspondence matrix $\mathbf{U}$ and its inverse $\mathbf{U}^{-1}$ given by

$$\mathbf{U} = \begin{pmatrix} 1 & 1 & 1 & 1 \\ 1 & 0 & 1 & 0 \\ 1 & 1 & 0 & 0 \\ 1 & 0 & 0 & 0 \end{pmatrix} \quad ; \quad \mathbf{U}^{-1} = \begin{pmatrix} 0 & 0 & 0 & 1 \\ 0 & 0 & 1 & \bar{1} \\ 0 & 1 & 0 & \bar{1} \\ 1 & \bar{1} & \bar{1} & 1 \end{pmatrix}$$

Here $\mathbf{U} = \mathbf{U}^T$ and hence $\mathbf{U}^{-1} = \mathbf{U}^{-1T}$. In addition $\mathbf{U}^{-1}$ can be obtained by inverting the order of the rows and of the columns of $\mathbf{U}$ and then reversing the signs of the second and third rows and columns. Hence $\mathbf{U} \equiv \mathbf{U}^{-1}$ and, therefore, $\mathbf{U}$ should generate a non-classical twinning mode of group 6. On solving equations (7) - (9) this mode is found to be

$$\{1110\} \; \{3112\} \; \langle 2\bar{1}\bar{1}3 \rangle \; \langle 011\bar{1} \rangle \; ; \; 5^{\frac{1}{2}}.$$

Note that the scalar product of K_1 and η_1 is zero indicating that the twinning direction lies in the twinning plane, and similarly for K_2 and η_2. In addition, ignoring signs, the same integers arise in K_1 and η_2 and in K_2 and η_1, which is a feature of this group in 3-d modes. However the Miller indices are here all rational, whereas all known examples of this group in 3-d have irrational indices.

Consider now the more general correspondence

$$\mathbf{U} = \begin{pmatrix} 1 & 0 & 0 & 0 \\ a & \bar{1} & 0 & 0 \\ b & 0 & \bar{1} & 0 \\ c & 0 & 0 & 1 \end{pmatrix} \quad ; \quad \mathbf{U}^{-1} = \begin{pmatrix} 1 & 0 & 0 & 0 \\ a & \bar{1} & 0 & 0 \\ b & 0 & \bar{1} & 0 \\ \bar{c} & 0 & 0 & 1 \end{pmatrix}$$

where a, b and c are any integers. Here $\mathbf{U} \neq \mathbf{U}^T$ but by changing the signs of the last row and last column it is seen that $\mathbf{U} \equiv \mathbf{U}^{-1}$. This is therefore a non-classical group 3 mode and equations (7) - (9) give

$$\{1000\} \; \{a^2 + b^2 + c^2, \bar{2}a, \bar{2}b, 2c\} \; \langle 0\bar{a}\bar{b}c \rangle \; \langle 2ab\bar{c} \rangle; \; (a^2 + b^2 + c^2)^{\frac{1}{2}}.$$

Again, η_1 lies in K_1 and η_2 in K_2 and all the Miller indices are rational. However, in keeping with a group 3 mode, the four Miller indices are distinct. The special case of this mode which gives the smallest possible shear g arises when one of a, b, c is unity and the other two are zero. For example a = 1, b = c = 0 gives

$$\{1000\} \; \{1\bar{2}00\} \; \langle 0\bar{1}00 \rangle \; \langle 2100 \rangle; \; 1.$$

It is interesting that this special case involves a degeneracy in which the same integers arise in K_1 and η_1 and in K_2 and η_2. The striking feature however is that because the third and fourth indices are all zero this is effectively a conventional 2-d twinning mode.

5. CONCLUSION

The classical theories of the crystallography of deformation twinning show that twinning modes are characterised by the elements K_1 K_2 η_1 η_2; g. Twinning shears can occur on either the twinning plane K_1, or the conjugate twinning plane K_2, but in each case the magnitude of the twinning shear is g. The modes may be Type I with K_1 and η_2 rational, Type II with K_2 and η_1 rational, or compound in which all four sets of Miller indices are rational. A generalised theory of deformation twinning due to Bevis and Crocker (1968) shows that modes can be divided into seven groups and only two of these have classical features. The remainder are in general characterised by four irrational twinning elements. This theory is formulated using the notation of the tensor calculus which readily enables twinning modes in 4-d lattices to be considered. In the present paper the case of the 4-d simple cubic lattice is considered and a few examples of possible twinning modes are presented. In these examples all of the twinning elements are rational, whereas the corresponding 3-d modes would have irrational elements. This suggests that twinning in 4-d may be a simpler process that in 3-d or even 2-d, modes being obtained by sectioning 4-d space. Severe problems of interpretation do however occur in 4-d. For example using the scalar product rule the plane (0001) contains the three perpendicular vectors [1000], [0100] and [0010] and therefore has the character of a 3-d volume.

Clearly more examples of 4-d twinning modes need to be determined in simple cubic, centred cubic and other 4-d lattices. Other crystallographic mechanisms can also be explored in 4-d. Indeed some progress has already been made in examining slip modes in 4-d (Crocker and Roberts 1991). One also wonders whether an extension of the Bullough and Bilby (1956) theory of martensite crystallography to 4-d might be worthwhile. Other themes of the present volume such as fracture and radiation damage might also be elucidated by considering 4-d analyses. However, discussion of these will have to await a future birthday celebration - perhaps when Ron Bullough reaches his century.

ACKNOWLEDGEMENTS

Many people over the past 35 years have contributed to the research on which this paper is based, particulary Bruce Bilby, Ron Bullough, Mike Bevis and more recently Elved Roberts. I am indebted to them all for their help and encouragement.

REFERENCES

Bacon DJ and Crocker AG 1966 Phil. Mag. **13**, 217
Bevis M and Crocker AG 1968 Proc. Roy. Soc. Lond. A **304**, 123
Bevis M and Crocker AG Proc. Roy. Soc. Lond. A **313**, 509
Bilby BA, Bullough R and Smith E 1955 Proc. Roy. Soc. Lond. A **231**, 263
Bilby BA and Crocker AG 1965 Proc. Roy. Soc. Lond. A **288**, 240
Bullough R 1957 Proc. Roy. Soc. Lond. A **241** 568
Bullough R and Bilby BA 1956 Proc. Phys. Soc. B **69**, 1276
Cahn RW 1954 Adv. Phys. **3**, 363
Crocker AG 1965 J. Nucl. Mat. **16**, 306
Crocker AG and Roberts E 1991 unpublished work
Guyoncourt DMM and Crocker AG 1968 Acta Metall. **16, **523.

Misfit dislocations

C J Humphreys

Department of Materials Science and Metallurgy, University of Cambridge, Pembroke Street, Cambridge CB2 3QZ.

ABSTRACT: The concept of a critical thickness for misfit dislocation introduction in lattice mismatched layers is discussed. Five possible heterogeneous sources for the introduction of the first misfit dislocations in an initially dislocation free substrate-epilayer system are identified. The characteristics of the "diamond-defect" source are described, and the possibility of growing dislocation free strained epilayers of large thickness is discussed.

1. INTRODUCTION

Seminal work on the dislocation content of boundaries with a general orientation relationship was performed over 35 years ago by Bilby, Bullough and Smith (1955), and further developed by Bullough (1964, 1965). These papers provide the framework for the study of misfit dislocations in mismatched epitaxial systems. It is widely believed that beyond a critical thickness, h_c, a strained epilayer will no longer grow coherently and misfit dislocations will be incorporated at the epilayer/substrate interface to relax the lattice mismatch strain. Van der Merwe (1963) argued that as a thin epitaxial layer grew coherently upon a substrate with different lattice parameter a critical thickness, h_c, would be reached at which it was energetically favourable to accommodate the lattice misfit using an array of dislocations rather than by increasing the elastic strain in the epilayer. This is clearly an equilibrium argument which does not take into account either the dislocation introduction mechanisms or any energy barriers to nucleating the misfit dislocations.

Matthews (1975) and Matthews et al (1976) provided a model for the introduction of misfit dislocations, based on the existence of dislocations already in the substrate and threading up to the substrate/epilayer interface. According to their theory, the critical thickness occurs when the epilayer stress becomes sufficient to cause the existing threading dislocations in the substrate to bend over at the interface and form misfit dislocations. This model is also an equilibrium model and Willis, Jain and Bullough (1990) have shown the exact equivalence of the equilibrium theories of Van der Merwe (1963) and Matthews (1975) and Matthews et al (1976). Jain et al (1991) have considered in detail interactions between the misfit dislocations and have shown that when these interactions are taken into account the critical thickness is always lower than the value of h_c derived from the theories of Van der Merwe (1963) or Matthews (1975). Willis et al (1991) have considered the driving force required for introducing the 'last' misfit dislocation to complete the periodic array, as distinct from the considerations later in this paper for the introduction of the 'first' misfit dislocation.

All of the above models (Van der Merwe 1963, Matthews 1975, and the improved model of Willis et al 1990) provide a good description of the behaviour of many systems (e.g. Kuk et al 1983). However it is evident that they cannot adequately explain h_c for the growth of epilayers on dislocation free substrates since they ignore the problem of dislocation nucleation. There have been many reports (e.g. People and Bean 1985) of experimental determinations of h_c for epilayer growth on low dislocation density semiconductor substrates in which the observed critical thickness is far greater than that predicted by the equilibrium theories (Van der Merwe, 1963; Matthews, 1975). This implies that the kinetics of dislocation nucleation and propagation are central to our understanding of epitaxial semiconductor systems. Before studying this in detail it is necessary to take a closer look at the experimental determination of critical thickness since, as Fritz (1987) has argued, the apparent critical thickness must depend strongly on the experimental technique which is used.

As an example of a strained layer system, GeSi/Si(100) is considered (Ge has a lattice parameter about 4% larger than that of Si). Recent work on this system has been comprehensively reviewed by Jain et al (1990). For high lattice parameter mismatch (>2%), the misfit dislocations are an orthogonal array of mainly edge type dislocations whereas at low mismatch (<2%) orthogonal bundles of 60° dislocations are formed (Kvam et al., 1988). Eaglesham et al (1988) have performed a detailed study of coherency breakdown in low mismatch GeSi/Si(100), grown by MBE at 550°C, using X-ray topography to probe the critical thickness. (X-ray topography is of course much more sensitive than electron microscopy to detecting low densities of dislocations. The minimum dislocation density detectable using electron microscopy is about 10^5 dislocations cm^{-2} whereas X-ray topography can detect a single dislocation in a specimen). Finite dislocation densities (in excess of 10^3 cm^{-2}) were found for an epilayer thickness a factor of 4 less than the accepted critical thickness, determined from TEM, RBS, or XRD, for this lattice mismatch. This result demonstrates that in a low-mismatched system the critical thickness h_c is not easily defined experimentally, since for a given epilayer thickness the dislocation density apparently increased continuously with increasing Ge content. There is not an abrupt change from no dislocations to some dislocations at a particular critical thickness: at very low dislocation density some regions of the specimen will have zero dislocations whereas other regions of the same specimen normally have a finite dislocation density. We conclude that, at least in low mismatched systems, there is no sharply defined critical thickness, and that misfit dislocations may exist at epilayer thicknesses substantially below the critical thickness reported in the literature.

2. THE SOURCE OF THE FIRST MISFIT DISLOCATIONS

For epilayer growth on a dislocation free substrate, what is the source of the first misfit dislocations? Frank (1950) and Hirth (1963) have shown that the lowest energy route is through the nucleation and propagation of a dislocation half-loop from the growth surface. Matthews etc al (1976) have considered in detail dislocation half-loop nucleation and propagation in strained epilayers and have given expressions for the critical radius a dislocation half-loop must have for it to propagate and the corresponding activation energy required. It was concluded that the nucleation barrier could not be overcome (at typical growth temperatures) for misfits below about 2% for *any* epilayer thickness. Eaglesham et al (1989a) have re-examined these calculations using a higher value of the dislocation core parameter (probably appropriate for dislocations in semiconductors) and calculate that the nucleation energy for a critical-radius half-loop is significantly higher than that calculated by Matthews: the new value is about 100 eV at 2% misfit, and it increases to about 1 000 eV as the misfit tends to zero. The very large nucleation energies required contrast with the experimental nucleation barrier of 0.7 eV measured by Hull et al (1988) to produce the observed temperature dependence of dislocation densities in 1.26% misfit GeSi/Si(100).

From the above discussion an interesting problem arises. The experimental evidence from X-ray topography (Eaglesham et al 1988) and electron microscopy (Eaglesham et al 1989b;

Hull et al 1988) is that it is rather easy to nucleate misfit dislocations in strained epilayers at low mismatch when the layers are grown on dislocation free substrates. The theoretical calculations (Eaglesham et al 1989a) however show that homogeneous nucleation of misfit dislocations should not be possible at any epilayer thickness for lattice mismatches of less than about 2%. It therefore seems that heterogeneous dislocation sources are readily available, even in relatively pure MBE grown material, unless very great care is taken.

Five such sources for the nucleation of misfit dislocations have been found so far for the GeSi/Si system: (i) regions of crystalline damage at the edges of GeSi/Si wafers (Tuppen and Gibbings 1990); (ii) internal precipitates in the epilayer (Tuppen and Gibbings 1989a and b); (iii) precipitates at the Si substrate/buffer interface due to inadequate substrate cleaning (e.g. carbon remaining at the substrate surface results in the formation of SiC precipitates, Perovic et al 1990); (iv) precipitates at the epilayer surface due to impurity contamination after growth (Higgs et al 1991); (v) "diamond defects", which are internal stacking faults in the epilayer (Eaglesham et al 1989a and b, Humphreys et al 1989, 1991). These five sources of dislocations give rise, respectively, to misfit dislocations originating (i) near the wafer edge, (ii) internally in the epilayer, (iii) at the substrate/buffer or, in the absence of a buffer layer, substrate/epilayer interface, (iv) internally in the epilayer, (v) at the epilayer surface. At least some of these heterogeneous dislocation sources are expected to occur in a wide range of lattice mismatched systems. In the case of the "diamond defect" sources, care was taken to observe well away from the edges of the wafer, to have good substrate cleaning and to avoid contamination after growth. Thus after eliminating (or minimising) nucleation sites (i) to (iv) above it was found that the diamond defect sources in our samples were the sources from which every misfit dislocation could be traced. This source has remarkable properties and is discussed further below.

The introduction of misfit dislocations in GeSi on Si(100) was studied using epilayers which were grown by MBE on deliberately unrotated substrates to provide graded compositions across the layers. Thus in addition to studying a critical thickness transition for epilayers of difference thickness at a fixed composition, it was also possible to study the same transition at a fixed thickness as a function of composition. Epilayers of nominal 20% Ge composition (0.8% lattice mismatch) were grown in bands of different thickness, and the source and substrate geometry gave a composition difference of typically ± 4 at % across the unrotated wafer. Further experimental details are given in Eaglesham et al (1989a and b).

At low mismatch, dislocation half-loops are observed and the misfit segments lie in orthogonal arrays (along [011] and [0$\bar{1}$1] for a (100) substrate). These misfit segments are predominantly 60° in character (~99%), very long (10-100 μm) and are not evenly spaced but grouped in bundles (Kvam et al 1988). Stereo microscopy and trace analysis show that the dislocation half-loops in a given bundle may all lie on the same inclined {111}, so in these cases the misfit dislocation segments are *not* coplanar with the epilayer/substrate (100) interface: the lowest misfit dislocation segment in a bundle lies at, or near, the heterointerface, with the other misfit segments lying in the epilayer on approximately the same inclined {111}. In addition, although a single bundle often consists of dislocation half-loops having only one $a/2\langle110\rangle$ Burgers vector, bundles are frequently populated with coplanar dislocation half-loops having two distinct Burgers vectors. (For example, a bundle of dislocations with misfit segments lying along [0$\bar{1}$1] may all lie on the same (1$\bar{1}$1) plane with some dislocations in the bundle having Burgers vector $a/2$[110] and some having $a/2$[101]. Both of these types of dislocations are glissile on (1$\bar{1}$1), the misfit segments being 60° type). Also a bundle may consist of dislocation half-loops lying on oppositely inclined {111}s and therefore populated with two distinct Burgers vectors (e.g. a bundle of dislocations with misfit segments along [011] may have some lying on the (1$\bar{1}$1) plane with Burgers vector $a/2$[110], and some on (11$\bar{1}$) with Burgers vector $a/2$[101]: both dislocations are glissile, on (1$\bar{1}$1) and (11$\bar{1}$) respectively, and the misfit segments are of 60° type).

This microstructure suggests that the bundles of dislocation half-loops may arise from a single source, capable of operating repetitively and, in particular, producing spatially correlated dislocations with different Burgers vectors on the same glide plane. All known regenerative sources (Frank-Read, double cross-slip and vacancy condensation) produce coplanar dislocations having the same Burgers vector. The source required to explain these observations therefore has unusual properties. Most of the 60° misfit dislocation segments observed in the material under investigation are so long (~50μm) that only a small percentage have their entire length located in the thin region of the crystal and these extended half-loops consistently exhibit the following properties: the misfit segment has 60° character and the two threading dislocations connecting this segment to the surface are usually screws, although 60° segments are also common.

3. CHARACTERISTICS OF THE DIAMOND DEFECT

Detailed studies of GeSi/Si epilayers from several different wafers consistently revealed the presence of diamond shaped planar faults (Eaglesham et al, 1989a and b), such as the one in the inset of figure 1. Diamond defects are not immediately obvious: the region of interest is very thick and the diffracting conditions change rapidly across the field-of-view because of the relatively high density of dislocations within the bundles and the bundle geometry. On the other hand, if the misfit dislocation density is very low, the presence of diamond defects can be mistaken for surface features arising from specimen preparation artifacts.

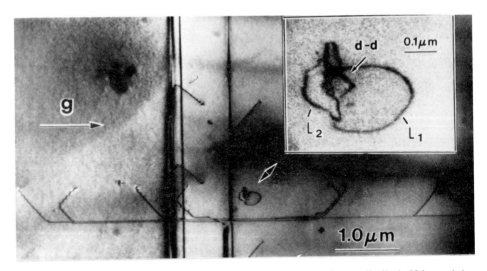

Fig.1. Bright-field 022 image showing a sequence of four coplanar glissile half-loops lying on $(11\bar{1})$ with mixed Burgers vectors (i.e. $\underline{b} = a/2[101]$ and $\underline{b} = a/2[1\bar{1}0]$). At the centre of the innermost half-loop there is a microstructural feature which is suggestive of a heterogeneous dislocation source. With reference to the inset: Diffraction contrast experiments show that the microstructural feature at the centre of the innermost half-loop consists of a diamond-shaped stacking fault which is out of contrast for the operating reflection, therefore only the diamond-shaped boundary dislocation (d-d) is visible, and that the diamond defect is associated with two glissile dislocation loops (L_1 lying on $(11\bar{1})$ with $\underline{b} = a/2[101]$, and L_2 lying on (111) and having $\underline{b} = a/2[1\bar{1}0]$).

Some characteristics of this new source are as follows. Diamond defects are typically 20 to 200 nm across and the number varies widely from 10^9 cm^{-3} to 10^{12} cm^{-3}. Stereomicroscopy, bright-field and weak-beam analyses have shown that diamond defects lie on {111} with inclined <110> edges, have a displacement vector of $a/6$<114> and the bounding dislocation image exhibits inside/outside behaviour which is consistent with a compressive state (i.e. interstitial in character) (Humphreys et al, 1989). The displacement vector $a/6$<114> is unusual but it has previously been reported. For example, $a/6$<114> faulted defects occur in silicon following ion implantation (Salisbury, 1982). These defects had the form of six-sided polyhedra lying on {111} with edges comprising the three <110> directions to be found in the {111} habit. The «missing» <100> edge in the four-sided diamond defect is the one perpendicular to the growth direction, and the Burgers vector of $a/6$<114> is perpendicular to this «missing» edge. The detailed morphology of the diamond defects is given in Humphreys et al (1989, 1991).

4. THE DIAMOND DEFECT AS A REGENERATIVE SOURCE

An $a/6[411]$ partial dislocation bounding the diamond shaped stacking fault on (111) may dissociate by one of the following reactions (Eaglesham et al, 1989b):

$$a/6[411] \rightarrow a/2[101] + a/6[11\bar{2}]$$
or
$$a/6[411] \rightarrow a/2[110] + a/6[1\bar{2}1]$$

Our observations are that the diamond defect always remains faulted. Hence instead of an unfaulting reaction occurring, the glissile $a/2$<110> dislocation segment bows out under the epilayer strain to form a half-loop attached to the $a/6$<112> partial dislocation at each end. The glissile $a/2$<110> dislocation then closes back on itself and recombines with the $a/6$<112> partial dislocation to leave a final configuration of the original $a/6$<114> diamond defect plus an $a/2$<110> dislocation loop propagating outwards on one of three possible {111}s. Hence the diamond defect (e.g. $a/6[411]/(111)$) can generate two types of glissile dislocations on the same glide plane (i.e. $a/2[110]$ and $a/2[101]$ on $(1\bar{1}1)$), as well as one glissile dislocation on each of two oppositely inclined glide planes (i.e. $a/2[110]$ on $(11\bar{1})$ and $a/2[101]$ on $(1\bar{1}1)$). If the diamond defect has an unique $a/6$<411> Burgers vector, then a given diamond defect cannot directly be the source of a glissile $a/2$<110> dislocation on its habit plane, as was suggested in Eaglesham et al (1989a) and mistakenly concluded in Eaglesham et al 1989b. The operation is identical to that of a Frank-Read source, and the diamond defect can emit $a/2$<110> dislocation loops repetitively.

The unique feature of the diamond defect is that it can generate $a/2$<110> dislocations with the same or with two different Burgers vectors, and that it can generate orthogonal bundles of misfit dislocation segments and epithreading dislocation segments whose properties are precisely those observed experimentally. The diamond defect is the only known source which can generate sequences of dislocation half-loops with different Burgers vectors. It seems clear therefore that the diamond defect is a new regenerative source of misfit dislocations in GeSi/Si material studied in the present investigations. Whether this source exists more widely in other materials systems requires further assessment. In particular it is important to understand all possible heterogeneous sources of misfit dislocations so that these can be controlled. If all such sources can be eliminated it should be possible to grow dislocation free low-mismatched strained epilayers to thicknesses very much greater than the equilibrium critical thickness.

REFERENCES

Bilby B A, Bullough R and Smith E *1955 Proc. Roy Soc.* **A.231** 263

Bullough R 1964 *Dislocations (AERE: Harwell)*

Bullough R 1965 *Phil. Mag.* **12** 1139

Eaglesham D J, Kvam E P, Maher D M, Humphreys C J and Bean J C, (1989a) *Philos Mag.* **59** 1059

Eaglesham D I, Kvam E P, Maher D M, Humphreys C J, Green G S, Tanner BK and Bean J C, (1988) *Appl. Phys. Lett.* **53** 2083

Eaglesham D J, Maher D M, Kvam E P, Bean J C and Humphreys C J (1989b), *Phys. Rev. Lett.* **62** 187

Frank F C, (1950) *Symposium on Plastic deformation of Crystalline Solids (Carnegie Inst. of Technology, Pittsburgh)* 89

Fritz I J, (1987) *Appl. Phys. Lett.* **51** 1080

Gomez A, Cockayne D J H, Hirsch P B and Vitek V, (1975) *Phil Mag.* **A 31** 105

Higgs V, Knightley P, Goodhew P J and Augustus PD, (1991) *Appl. Phys. Lett.* **59** 829

Hirth J D, (1963) *In Relation Between Structure and Strength in Metals and Alloys (HMSO: London)* 218

Hull R, Bean J C, Wader D J and Leibenguth R E, (1988) *Apply Phys. Lett.* **52** 1605

Humphreys C J, Maher D M, Eaglesham D J, Kvam EP and Salisbury I G, (1991) *J. Phys. III* **1** 1119

Humphreys C J, Maher DM, Eaglesham DJ and Salisbury I G, (1989) *Inst. Phys. Conf. Ser. No. 100* 241

Humphreys C J, Eaglesham D J, Maher D M, Fraser H L and Salisbury I G, (1990) *in Evaluation of Advanced Semiconductor Materials by Electron Microscopy, Ed. D. Cherns (Plenum Press: London and New York)* 203

Jain S C, Gosling T J, Willis J R, Totterdell D H J and Bullough R, (1991) *Phil. Mag (in press)*

Jain S C, Willis J R and Bullough R, (1990) *Adv. in Physics* **39** 127

Kuk Y, Feldman L C and Silverman P J, (1983) *Phys. Rev. Lett.* **50** 511

Kvam E P, Eaglesham D J, Maher D M, Humphreys C J, Bean J C, Green G D and Tanner B K, (1988) *Mat. Res. Soc. Symp. Proc.* **104** 623

Matthews J W, (1975) *J. Vac. Sci. Technol.* **12** 126

Matthews J W, Blakeslee A E and Mader S., (1976) *Thin Solid Films* **33** 253

People R and Bean J C, (1985) *Appl. Phys. Lett.* **47** 327

Perovic D D, Weatherly G C, Baribeau J-M and Houghton D C, (1990), *Thin Solid Films (in press)*

Salisbury I G, (1982) *Acta. Metall.* **30** 27

Tuppen C G and Gibbings C J, (1990) *Thin Solid Fims (in press)*

Tuppen C G, Gibbins C J and Hockly M, (1989a) *J. Cryst. Growth* **94** 392

Tuppen C G, Gibbings C J and Hockly M, (1989b) *Mat. Res. Soc. Symp. Proc.* **130 141**

Van der Merwe J H, (1963) *J. Appl. Phys.* **34** 123

Willis J R, Jain S C and Bullough R (1990) *Phil. Mag.* **A62** 115

Willis J R, Jain S C and Bullough R, (1991) *Appl. Phys. Lett. (in press)*

Modelling dislocation structure and mobility in boundaries in h.c.p. metals

D J Bacon[1] and A Serra[2]

[1] Department of Materials Science and Engineering, University of Liverpool, P.O. Box 147, Liverpool, L69 3BX, U.K.

[2] Departament de Matematica, Aplicada III, Universitat Politecnica de Catalunya, ETSE Camins, Gran Capitan s/n, 08034 Barcelona, Spain

ABSTRACT: Computer simulation of the atomic structure and movement of twinning dislocations in h.c.p. metals is reviewed. These dislocations have the form of steps on the twin boundary, and whereas some have cores which are very widely spread over the interface, others are only an interatomic spacing or so across. These configurations are determined mainly by the complexity of any shuffles required to restore the crystal structure when the dislocation is introduced. The mobility of the dislocations is also controlled by the same effect, and is found to correlate well with experiment.

1. INTRODUCTION

The properties of crystal interfaces has long been of interest, but what has distinguished recent work from earlier studies in the field has been the contribution of computer simulation to our understanding of interface structure at the atomic scale. This has tended to be concerned with the structural arrangements of atoms, and has concentrated on the cubic metals. Very recently, this branch of modelling has started to be applied to the lower symmetry hexagonal-close-packed (h.c.p.) structure,and the purpose of the present paper is to explain how this has provided insight into the movement of dislocations in boundaries. This is an important problem for the h.c.p. metals because one of their major plastic-deformation mechanisms is deformation twinning (Yoo 1981), and this is believed to involve motion of a twinning dislocation along the twin boundary. The h.c.p. metals exhibit a variety of twinning modes, and the dislocations that can exist in the interfaces for these modes are different, so that a spectrum of dislocation structure and behaviour can be studied.

A twinning dislocation has the form of a step in the twin-matrix interface and glides along this boundary in response to a shear stress resolved in the twin interface plane K_1 in the direction of the twinning shear η_1.For a particular twin plane, several twin modes are possible, and each has a distinct Burgers vector b and step height h for its twinning dislocation. (A full discussion of the analysis from which defects in h.c.p. twin boundaries may be specified is presented in Serra et al. (1988,1991).) The twins which have been modelled to date have K_1 equal to $\{10\bar{1}1\},\{10\bar{1}2\},\{11\bar{2}1\}$ or $\{11\bar{2}2\}$. They are widely observed in the h.c.p. metals, and their crystallographic elements have been determined by experiment. Furthermore, the extent to which a particular mode occurs varies from metal to metal and depends on the load and temperature conditions. Twinning dislocations in these boundaries are therefore a natural choice for theoretical investigation.This paper reviews current understanding of this area, based on the simulation studies of Serra et al. (1988,1991), Pond et al. (1991) and Serra and Bacon (1991a,1991b).

2. METHOD

A schematic representation of a twinned crystal containing a twinning dislocation in the edge orientation is drawn in Figure 1.The plane of shear is the x-z plane, and the conjugate twinning plane and direction, K_2 and η_2, are as indicated. The sense of the step is reversed if the sign of either the dislocationor η_1 is reversed. Planar, dislocation-free interfaces were first modelled for each twin to determine the boundary configuration of minimum energy, taking care to ensure that all possible interfacial structures were sampled in this process. Next, a twinning dislocation was introduced at the interface (as in Figure 1) and the atoms allowed to relax.The inner region of the computational block contained up to 1000 atoms, interacting by one of several different potentials. In early work (Serra and Bacon 1986,Serra et al. 1988,1991), pair potentials were used, among which that named na56 has been argued to be one of the best for the h.c.p. structure. This oscillatory function offers a good representation of a near-ideal c/a ratio metal such as magnesium (Mg). Very recently, Serra and Bacon (1991a,1991b) have employed many-body potentials, which are better able to describe the elastic constants, cohesive energy and c/a ratio of real metals. The potentials used to date are those of Igarashi et al. (1991) for Mg and Ackland (1991) for titanium (Ti).

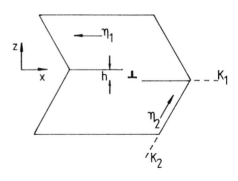

Figure 1. Illustration of a twinning dislocation

3. DISLOCATION CORE STRUCTURE

As noted above, many different dislocations may result in twinning on a twin plane K_1, each corresponding to a unique set of elements η_1, K_2 and η_2, but in practice only one (or rarely two) seems to occur. It has long been argued (e.g.Crocker and Bevis (1970)) that the mode that actually operates should have a low magnitude of shear s $(=b/h)$. In our analysis we have therefore concentrated on those with small s and b. In the following, the step height associated with a dislocation core will be indicated by a subscript on b: thus, for b_n the step height is n times $d(K_1)$, the spacing of the K_1 lattice planes.

Consider first the $(11\bar{2}1)$ boundary. Only one twinning dislocation meets the criterion specified above, namely $b_{1/2}=1/70[11\bar{2}6]$. (Here and subsequently, the quotient preceding the direction indices of b depends on c/a and is given in approximate form for $c/a =\sqrt{8/3}$.) The core configuration for Mg is plotted in Figure 2a. The lattice translation vectors t and basis vectors w are those whose differences define b (see Serra et al. 1988). The hydrostatic component p of stress at each atomic site is mapped in order to indicate both the stress state in the interface and core and the shape of the interfacial step. (The scale mark for p is in units of Boltzmann's constant k times the absolute melting temperature T_m divided by the volume per atom Ω in the perfect crystal.) It can be seen that the step is very broad and has a half-width spreading over a region some $10a$ across, i.e. about $40b$.This dislocation also has a very wide form in the na56 and Ti models.

Only one dislocation need be considered for the $(10\bar{1}2)$ twin. It has $b_2=1/17[10\bar{1}\bar{1}]$ and is plotted for the Mg potential in Figure 2b. Here, atoms in the lattice plane between the two interface planes have to 'shuffle' in order to reach their correct positions, but these movements are relatively simple to achieve. Although the step is considerably higher than that for the $b_{1/2}$ dislocation in the $(11\bar{2}1)$ twin, the core is seen to be wide, as it is also in the na56 and Ti models, and has a half-width of approximately $6a$, again about $40b$.

The situation for the $(11\bar{2}2)$ boundary is more complex, for several dislocations are suitable, in principle, for consideration under the guidelines stated above. The one with the lowest h has $b_1=1/11[11\bar{2}3]$. The core is wide, but the shear is relatively large in magnitude and has the reverse sense to that observed experimentally. The next highest step and low s is created by a dislocation with $b_3=2/33[11\bar{2}3]$. This requires that atoms in two intermediate planes shuffle to restore correct structures, and this can be achieved in several different ways. These have been investigated (Serra et al. 1988) and the core state with the lowest energy is plotted for the na56 model in Figure 2c.The b_3 dislocations in Mg and Ti are similar to this. It can be seen that this core is strongly localised with large compressive stresses, and occupies a region only 2-3a across. Another $(11\bar{2}2)$ dislocation is theoretically possible with an even lower s. It has

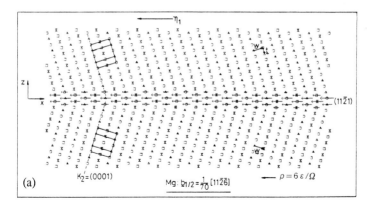

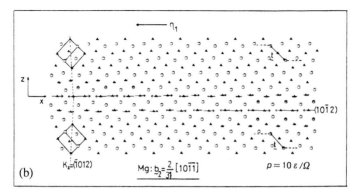

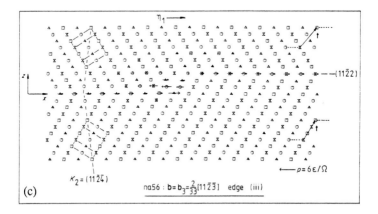

Figure 2 (Continued over)

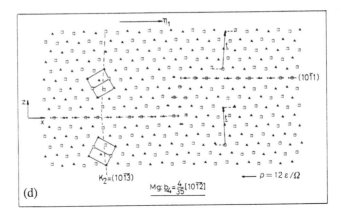

(d)

$K_2=(10\bar{1}3)$ $Mg: b_4=\frac{4}{35}[10\bar{1}2]$ $p=12\,\varepsilon/\Omega$

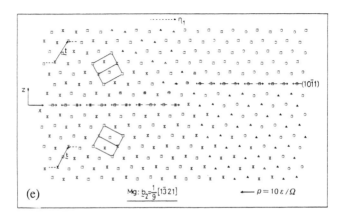

(e) $Mg: b_2=\frac{1}{9}[1\bar{3}21]$ $p=10\,\varepsilon/\Omega$

Figure 2.Projections along the line of the relaxed atomic structure around the cores of twinning dislocations for (a) $b_{1/2}$ in the $(11\bar{2}1)$ twin, (b) b_2 in the $(10\bar{1}2)$ twin, (c) b_3 in the $(11\bar{2}2)$ twin, and (d) b_4 and (e) b_2 in the $(10\bar{1}1)$ twin. (c) is for na56, the others for Mg.

$b_4=1/33[11\bar{2}\bar{3}]$ and $h=4d(K_1)$, and therefore requires shuffles on three intermediate planes of the step. The various core states resulting from this were investigated for the na56 potential by Serra et al. and none was as energetically favourable as the b_3 dislocation described above.

Twinning on $(10\bar{1}1)$ can arise from one of three dislocations corresponding to relatively low magnitude of shear s. They have step heights of two, three or four times $d(K_1)$ and thus require shuffling on one, two or three lattice planes respectively. The dislocation with b_3 has the largest s and cores with the largest distortion and highest energy. The lowest energy state for the $b_4=5/41[\bar{1}012]$ dislocation is plotted for Mg in Figure 2d. The step is very narrow but does not produce significant distortions. The b_3 and b_4 dislocations have the same direction of b and hence η_1.The third $(10\bar{1}1)$ dislocation has an intermediate value of s, but a step height of only $2d(K_1)$. It gives rise to type I twinning with η_1 and K_2 both irrational. For c/a close to ideal, b_2 is approximately $1/9[1\bar{3}21]$. Since b is irrational, it is not possible to simulate the pure edge and screw forms of this dislocation, and so Serra et al. considered the mixed line in the same orientation as that of the line in Figure 2d, and combined edge and screw components of approximately $1/18[\bar{1}012]$ and $1/6[1\bar{2}10]$ respectively. The atomic structure of this dislocation in the Mg is plotted in Figure 2e, from which it seen that the core is again narrow and relatively free of stress. The core structure found for these dislocations in na56 and Ti is similar to that shown here for Mg.

4. DISLOCATION MOVEMENT

Glide of a twinning dislocation was simulated by displacing all the atoms in response to a simple shear strain $e=e_{xz}$, which was increased in increments of 0.001 or more, with relaxation after each increment, until free dislocation movement occurred at $e=e_c$, the critical strain. For the type I (10$\bar{1}$1) mode e comprised both e_{xz} and e_{yz} in appropriate proportions.

It was found that the mobility of the twinning dislocations as measured by e_c falls into three categories. Some dislocations move at strains of the order of the lowest value applied, i.e.~0.001, whereas others have e_c values an order of magnitude higher at about 0.01. The third type could not be induced to glide even for applied strain values of ~0.02, and in some cases did not move even for $e=0.04$ The actual e_c values depend slightly on the choice of interatomic potential, but this general classification applies to all the model metals. The highly mobile dislocations are those with $b_{1/2}$ in the (11$\bar{2}$1) boundary, b_1 in the (11$\bar{2}$2) and b_2 in the (10$\bar{1}$2), and it may be noted that these three are those with the widest cores. The intermediate behaviour was exhibited by the b_2 dislocation in the (10$\bar{1}$1) interface.The b_4 cores in the (10$\bar{1}$1) and (11$\bar{2}$2) twins and the b_3 dislocation in the (11$\bar{2}$2) boundary were found to be essentially sessile with all potentials, although the latter was induced to move at high strain in the na56 crystal. It will be observed that these dislocations of low mobility are those with cores in the form of narrow, abrupt steps in the interface.

5. DISCUSSION

The results reviewed here are the most extensive obtained to date on the h.c.p. metals, and they demonstrate clear differences in core morphology and mobility between the twinning dislocations of the (11$\bar{2}$1) and (10$\bar{1}$2) twins on the one hand and those of the (11$\bar{2}$2) and (10$\bar{1}$1) twins on the other. It should be noted that these characteristics are qualitatively the same in all three potential models. The results are summarized in Table 1, where data are included for the twin boundary energy γ, the dislocation line energy and the contribution to it of the core energy. γ is lowest in every case for the Ti model and highest for Mg, and, with the exception of the (10$\bar{1}$2) interface, there is little to choose between the twins. The line energy is only approximate because the x-z cross-sectional area of the computational model was not the same in each case. Nevertheless, the differences between the different dislocations is striking. Furthermore, when the elastic contribution to the energy is estimated and subtracted (Serra et al. 1991), the remaining core term shows a wide (and more meaningful) variation.

Table 1. Twin-boundary energy γ (units kT_m/a^2), and line energy (units kT_m/a) and other parameters for the twins listed. (* this dislocation simulated with na56 potential only)

Twin	γ	b	s	b^2/a^2	$h/d(K_1)$	Line energy	Core energy	Core width	e_c
(10$\bar{1}$1)	0.6-1.1	b_2	0.36	2/7	2	2.8-5.3	0.4-1.9	Narrow	Intermediate
		b_4	0.15	25/123	4	2.6-5.5	1.3-2.9	Narrow	High
(10$\bar{1}$2)	0.9-1.5	b_2	0.12	1/51	2	0.1-1.0	0.0-0.3	Wide	Low
(11$\bar{2}$1)	0.5-1.2	$b_{1/2}$	0.61	3/140	1/2	0.2-0.5	0.0	Wide	Low
(11$\bar{2}$2)	0.7-1.2	b_1	1.22	3/11	1	3.1*	0.1*	Wide*	Low*
		b_3	0.27	4/33	3	3.7-6.8	2.1-4.6	Narrow	High

The Table demonstrates very clearly that for dislocations in which the atomic disregistry is localized into a narrow step, the core energy is an order of magnitude larger than those in which it can spread easily over the interface. The ability of the core to spread is not determined by γ, nor is it simply related to the step height h, for the b_2 dislocation in the $(10\bar{1}2)$ boundary has a wide core. It would appear to be dependent on whether or not complex atomic shuffles are required, in addition to the displacement b, to restore correct coordination. No shuffles are necessary for the $b_{1/2}$ dislocation in $(11\bar{2}1)$ nor b_1 in $(11\bar{2}2)$, and the shuffles are relatively simple for the b_2 line in $(10\bar{1}2)$. For the other cases, however, the shuffles are more difficult to achieve, and this is believed to lead to the abrupt steps and large core energy terms. The possible importance of shuffles for twinning in the h.c.p. metals was recognised some 20 to 30 years ago (e.g. Crocker and Bevis 1970), and the computer simulations underline this in terms of the way in which shuffles restrict the spread of core structure along the interface. The results discussed here also emphasize the correlation between dislocation mobility and core width. The wide cores are able to glide along their interface under very low shear stress, whereas the abrupt steps are very difficult, or even impossible, to move. In fact, attempts to induce the b_3 dislocation in the $(11\bar{2}2)$ boundary to glide by imposing shuffle-like displacements in the core whilst applying the shear strain failed, even for e as large as 0.03-0.04 (Serra et al. 1991).

Finally, the simulations described here should be put into the context of experimental information on twinning in the h.c.p. metals. One would anticipate that the $(10\bar{1}2)$ and $(11\bar{2}1)$ modes should propagate easily, and, indeed, these two twins are widely reported under conditions of low stress and low-to-moderate temperature (see discussion in Serra et al. 1991). They accommodate shears imposed by tension along the [0001] direction. So does the b_1 dislocation in the $(11\bar{2}2)$ boundary, yet this mode is not observed experimentally. We assume that this twin, although having a mobile dislocation, is unable to compete with the $(10\bar{1}2)$ and $(11\bar{2}1)$ twins because of the combination of its large shear and relatively high line energy. The other three modes listed in the Table have senses of shear which would arise under [0001] compression. All are associated with twinning dislocations which are difficult to move, and this is entirely consistent with the experimental finding that $(10\bar{1}1)$ and $(11\bar{2}2)$ twinning only occurs under conditions of high stress and temperature. However, the $(10\bar{1}1)$ mode actually observed (Reed-Hill 1960, Paton and Backofen 1969) corresponds to the b_4 dislocation, not the b_2 which, according to our findings, has a lower energy and a slightly higher mobility. The reason for this inconsistency is not clear. It may be that the experimental loading conditions favoured the b_4 system, or that twin nucleation, affected by the size of the twinning shear rather than ease of propagation, may be decisive for $(10\bar{1}1)$ twinning. This requires further investigation.

In conclusion, these simulations have revealed a wide range of structural characteristics and behaviour of dislocations in interfaces in the h.c.p. metals, and have provided a clear demonstration that core structure is the predominant factor in controlling interface-step mobility. This is in generally good agreement with interpretations based on experiments.

REFERENCES

Ackland G, 1991, unpublished work
Crocker A G and Bevis M, 1970, *The Science, Technology and Application of Titanium* (Ed. R.Jaffe and N Promisel), (Oxford, Pergamon Press), p.453.
Igarashi M, Khantha M, and Vitek V, 1991,*Phil Mag.* A, **63**, 603.
Paton N E and Backofen W A, 1969, *Trans.* AIME, **245**, 1369.
Pond R C, Bacon D J, Serra A and Sutton A P, 1991,*Metall. Trans.,* **22**A, 1185.
Reed-Hill R E, 1960, *Trans.* AIME, **218**, 554.
Serra A and Bacon D J, 1991a, *Phil. Mag.* A, **63**, 1001-1012.
Serra A and Bacon D J, 1991b, *MRS Sympos.on Structure and Properties of Interfaces*, Dec. 1991,to be published
Serra A, Bacon D J and Pond R C, 1988, *Acta Metall.*, **36**, 3183.
Serra A, Pond R C and Bacon D J, 1991 *Acta Metall. Mater.*, **39**, 1469.
Yoo M H, 1981,*Metall. Trans.* A, **12** 409.

Stable dislocation arrays in multilayer structures

J.P. Hirth

Department of Mechanical and Materials Engineering

Washington State University, Pullman, WA 99164-2920, USA

ABSTRACT

The stability of misfit dislocation arrays in multilayer structures is considered. The range of critical thickness is shown to be narrow for the stability of arrays of varying spacing at a fixed degree of misfit. The results are consistent with trends in experimental observations.

1. PREFACE

It is a pleasure for me to join in honouring Ron Bullough. Over the years, I have valued greatly the many discussions with him of aspects of dislocation theory and about the interactions of dislocations and point defects. The topic of the present note is one for which he has made significant contributions and for which he is actively engaged in on-going research.

2. INTRODUCTION

The observed critical thickness at which a coherently strained multilayer undergoes the injection of misfit dislocations is known to decrease with increasing misfit (Matthews, 1979, and People and Bean, 1985, 1986). In part, the trend is associated with kinematic constraints, such as a nucleation barrier to dislocation formations (Hirth and Evans, 1986 and Kamat and Hirth 1990), insufficient driving force for dislocation multiplication and motion Dodson and Tao, 1988, Dregia and Hirth, 1991, and Jain et al 1991, and possible Peierls barrier effects. For the often studied case of layers of *fcc* or diamond cubic films with {100} interfaces, there is also the complication that the dislocations that form at low misfit are the inefficient 60° dislocations rather than the Lomer dislocations that are most effective in removing misfit strains. However, the trend is also for the equilibrium critical thickness for a given misfit dislocation spacing to decrease with increasing misfit. The objective of the present work is to define the upper and lower bounds for such an equilibrium spacing.

Matthews and Blakeslee, 1974, 1975, calculated the critical thickness for the energies for misfit dislocation arrays using an atomistic piecewise parabolic type of potential. We have calculated interaction energies for arrays of Volterra dislocations and find agreement with the result of Van der Merwe and Jesser (1988) for the square array assumed by then, although we showed that such a square array is not the minimum energy array. The method involves an explicit expression for the interaction energy of two parallel dislocation arrays that is then summed for the multilayer case. Both of the above results, (Van der Merwe and Jesser 1988 and Hirth and Feng 1990) are for the case where the elastic constants of the two constituents phases in the multilayer are assumed to be the same. Willis et al (1990) have solved for the field of arrays of parallel dislocations by Fourier analysis, including effects of elastic inhomogeneity.

The physical nature of the problem is contained in the simpler trilayer case where analytical expressions are available for the stress fields and interaction energies, so we restrict the present treatment to that case. Also, the elastic constants of the constituent phases are assumed to be the same.

3. DISLOCATION ENERGIES

We consider a layer A contained within two layers B of much larger extent for two fcc or diamond cubic crystals joined along (001) interfaces as shown in Figure 1. The stress field in A is give by

$$\sigma = 2\mu\kappa\varepsilon = c\varepsilon \tag{1}$$

where μ is the stress modulus, $\mu = (1+v)/(1 - v)$, v is Poisson's ratio, and E is the coherency strain in layer A. For the trilayer case of Figure 1, $\varepsilon_0 = (a - a_0)/a_0$,

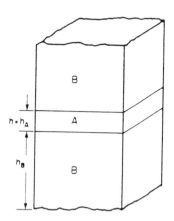

Fig. 1. A trilayer of cubic crystals

where a is the in-plane lattice parameter of the strained layer and a_0 is its unstrained equivalent. When misfit dislocations completely accommodate the misfit, the misfit dislocation spacing is related to ε_0 as

$$\lambda_0 = b/\varepsilon_0 \qquad (2)$$

where b is the edge component of the misfit dislocations. We treat the case where the interface dislocations are Lomer dislocations, pure edge dislocations with Burgers vectors $1/2[110]$ and $1/2[1\bar{1}0]$ laying in the interface in two orthogonal arrays. The extent L of the layers is considered to be so large compared to the thickness A layer thickness h that edge effects are negligible, Hirth and Evans (1986).

A reversible path for creation of the dislocations is shown in Figure 2. A

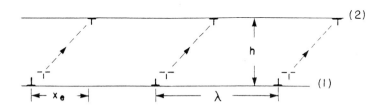

Fig. 2. Reversible creation of a set of dislocation multipoles

set of infinitesimal dislocation dipoles is created and array 2 is separated from array 1. Since the spacing within each array remains constant, there is no work done from dislocation interactions within either array. The work is the sum over the total number of dipoles of the interaction of a single dislocation in array 2 with the set in array 1 as separation occurs. Only the mean stress for the orthogonal array enters the problem, and this is accounted for in the factor κ in eq. (1). The net reversible work, which equals the interaction energy, is determined by the integration of the Peach-Koehler force over the separation path. Since this force contains the stresses of all the dislocations and the coherency stress, which can be superposed according to the principle of superposition, the result is unambiguous. No concerns with cross-terms, as arise if one superposes energies as discussed by Willis et al. (1990) appear.

The result for the total energy of an array of arbitrary spacing λ is

$$\frac{W}{L} = \frac{\mu b^2}{4\pi(1-\nu)}\left[\ln\left(\frac{2ph}{l}\right)\sinh\left(\frac{2\pi h}{l}\right) - \ln\left(\cosh\left(\frac{2\pi b}{l}\right)-1\right) - \frac{8\pi h}{\lambda_0}\right] \qquad (3)$$

where the dislocation core - radius has been set equal to b. Also in the total core energy, a core-type term $-\mu b^2/\pi(1-\nu)$ has been dropped so that the limiting form of W/L agrees with the single dipole result. The last term in eq. (3) is the contribution from the coherency strain and can alternatively be written as $(8\pi h\varepsilon_0/b)$. For a fixed misfit, ε_0, the variation of critical thickness h, as a function of A is determined from eq. (3) for the case that $(W/L) = 0$.

The equilibrium position x_3 in Figure 3 is determined from the condition.

$$\cos\left[\frac{2\pi x_e}{\lambda}\right] = \cosh\left[\frac{2\pi h}{\lambda}\right] - \left[\frac{2\pi h}{\lambda}\right]\sinh\left[\frac{2\pi h}{\lambda}\right]$$ (4)

4. DISCUSSION

For the particular case that $\lambda = \lambda_0$, the critical thickness h_0 is that for which the dislocation array completely removes the coherency strain, curve I in Figure 3. This is the case previously considered. In the limit that $\lambda \to$ infinity, $x_e = h$ and eq. (3) reduces to

$$\frac{W}{L} = \frac{\mu b^2}{4\pi(1-v)}\left[\ln\left[\frac{\sqrt{2}h}{b}\right] - \frac{4\pi h}{\lambda_0}\right]$$ (5)

which is the correct form for the Matthews-Blakeslee result for the triple layer case.

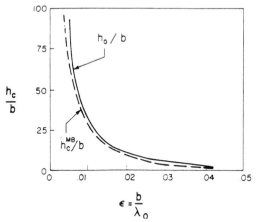

Fig. 3. A plot of h_c/b versus b/λ_0 for the trilayer case.

The value of h_c^{MB} determined for this limiting condition is also shown in Figure 3. As has been noted by Jain et al, if h is increased for a fixed value of ε_0 in eq. 3, in the coherent region, $h < h_c^{MB}$, the total energy increases monotonically with h. The total energy is a maximum at $h = h_c^{MB}$ and then decreases monotonically with h to h_0, above which the misfit strain is completely relaxed and the total energy is constant for the present case (in the multilayer case the total energy would continue to decrease with increasing h because larger h connotes fewer total misfit dislocations). Thus, the two curves in Figure 3 represent upper and lower bonds for possible equilibrium dislocation arrays. The range so defined is seen to be quite narrow.

In the range between the two bounds, as first discussed by Van der Merwe and Jesser (1988), as h $= h_c$ increases at a fixed misfit ε_0, the equilibrium value of λ decreases monotonically from infinity to λ_0. The practical consequence of this result is that in the absence of kinematic constraints, the first misfit dislocation should appear at h_c^{MB} but should not multiply. As h $= h_c$ increases, multiplication is favoured, but complete misfit removal should not occur. Only when h $\geq h_0$ should the spacing equal λ_0 and should the misfit be completely removed.

As mentioned previously, the same considerations apply qualitatively to the multilayer case. However, the sums required to qualitatively discuss this case, given in Hirth and Feng (1990), are too lengthy to introduce here. The result equivalent to Figure 3 for the multilayer case also reveals a narrow range for $h_c^{MB} < h_c < h_0$.

5. SUMMARY

In conclusion, the different treatments of the multiple misfit dislocation arrays are shown to give consistent results when the arrays that are treated have the same configuration. The single-dipole Matthews-Blakeslee result represents the lower bound in layer thickness for the presence of misfit dislocations at local equilibrium. The upper bound in layer thickness is that for which the total misfit strain is removed by an equilibrium array of misfit dislocations of spacing λ_0. Between these bounds the misfit dislocation spacing monotonically changes from infinity to λ_0.

6. ACKNOWLEDGEMENT

The author is grateful for the support of the research by the Defense Advanced Research Projects Agency University Research Initiative at the University of California, Santa Barbara under Office of Naval Research Contract No. N-00014-86-K-0753. The author is grateful to J.R. Willis for pointing out an error in an earlier form of eqs. (3) and (4) where the factor K was used instead of $1/(1 - v)$. The reduced results in ref 11 are not affected by this change and hence are correct.

7. REFERENCES

Dodson B.W. and Tsao J Y 1988 Phys. Rev. B **38** 12
Dregia S A and Hirth J P 1991 J. Appl Phys. **69** 2169
Hirth J P and Evans A G 1986 J. Appl. Phys. **60** 2372
Hirth J P and Feng X J 1990 Appl. Phys. **67** 3343
Hull R and Bean J C 1989 J. Vac. Sci. Technol. A **71** 2580

Jain S C, Gosling T J, Willis J R, Totterdell D H J and Bullough R 1991 Philos. Mag A in press.

Kamat S V and Hirth J P 1990 J. Appl. Phys. **67** 6844

Matthews J W 1979 Dislocations in Solids, edited by F.R.N. Nabarro (North-Holland Amsterdam) Vol. 2 pp. 461.

Matthews J W and Blakeslee A E 1974 J. Cryst. Growth. **27**, 118; ibid. 1975 **29** 273, 1976 ibid. **32** 265

People R and Bean J C 1985 Appl. Phys. Lett. **47** 322; ibid. 1985 **49** 229

Van der Merwe J H and Jesser W A 1988 J. Appl. Phys. **63** 1509

Willis J R, Jain S C, and Bullough R 1990 Phil. Mag. A **62** 115; in press.

Stability of capped Ge_xSi_{1-x} strained epilayers

S.C. Jain[1], T.J. Gosling[1,2], J.R.Willis[3], R. Bullough[2], P. Balk[1]
1. DIMES, T.U. Delft, Postbus 5053, 2600 GB Delft, NL., 2. A.E.A. Technology,
Harwell Laboratory, OXON OX11 ORA, U.K., 3 School of Mathematical Sciences,
University of Bath, BA2 7AY, U.K.

ABSTRACT: Critical layer thickness and strain relaxation in strained capped Ge_xSi_{1-x} epilayers are calculated and found to be in good agreement with experimental values.

1. INTRODUCTION

Strained semiconductor epilayers are playing an increasingly important role in the design and fabrication of advanced devices (Jain et al 1990, Jain and Hayes 1991). As a strained layer is being grown, its stability characteristics are those of a layer with a free surface. However, the completed devices usually have Si epilayers on both sides of the strained Ge_xSi_{1-x} layer (Jain et al 1990, Garone et al 1990, Patton et al 1990, Kesan et al 1990). As distinguished from the layers with a free surface, these strained layers are called capped layers; it is this configuration which determines the stability of the devices against strain relaxation by introduction of misfit dislocations (Houghton et al 1989, 1990, Houghton 1990a, b). Earlier authors (Houghton 1990a, b) have interpreted experimental results using a theory based on the result of the Matthews and Blakeslee (1974). It will be shown elsewhere (Gosling et al 1992) that the Matthews and Blakeslee model is not correct. In this paper, we use a theory which correctly takes into account interactions between dislocations, and between homogeneous strain and dislocations.

2. DISLOCATION MORPHOLOGY AND DISLOCATION ENERGIES

The capped layers considered here are Ge_xSi_{1-x} layers on (001)Si substrates with fractional mismatch parameter f_m. Strain relaxation is assumed to occur by the introduction of orthogonal arrays of dislocation dipoles (Willis et al 1991). The dislocations on the lower interface therefore have Burgers vector **b** and those on the upper interface have Burgers vector **-b** (Willis et al 1991). The dislocations are assumed to be of the 60° type (Matthews et al 1974). The spacing p is the separation between adjacent dipoles in the periodic array.

The strain in a layer may be split into a uniform strain, which has the value of the mean strain in the layer, and a fluctuating strain which has a mean value zero in the layer (Willis et al 1990, 1991). The total energy of the layer (Willis et al 1990, 1991) can be written as,

$$E_T = E_H + E_D. \tag{1}$$

E_H is the energy contribution from the *relaxed* misfit strain, which is the mean strain in the partially relaxed layer, and E_D is the contribution from the fluctuating part of the strain field associated with the array of dipoles. Other symbols are defined in the Appendix. Since we consider orthogonal periodic arrays, the relaxed misfit strain is given by (Jain et al 1990)

$$\varepsilon_r = f_m + \frac{b_1}{p}. \tag{2}$$

Here b_1 is the 'active' component of the Burgers vector, which tends to relax the misfit strain. For a given dislocation spacing, the relaxed misfit strain is the same in capped and uncapped layers (Willis et al 1990, 1991). The energy contribution per unit area from the relaxed misfit strain is given by

$$E_H = 2\mu \frac{1+\nu}{1-\nu} \varepsilon_r^2 h. \tag{3}$$

The energy contribution per unit area, E_D, from the fluctuating part of the strain field is given by (Willis et al 1990, 1991)

$$E_D = \frac{2}{p} E_{DS}. \tag{4}$$

E_{DS} is energy per unit length of a dislocation line due to the fluctuating strains. For an uncapped layer (Willis et al 1990, see correction in Jain et al 1992)

$$E_{DSu} = \frac{\mu}{4\pi(1-\nu)} \left[a_0 + a_1 \ln\left\{ p \frac{(1-e^{-s})}{2\pi q} \right\} + a_2 \frac{se^{-s}}{(1-e^{-s})} - a_3 \frac{s^2 e^{-s}}{(1-e^{-s})^2} - a_2 \right] \tag{5}$$

with $s = \left(4\pi \frac{h}{p} \right)$. The corresponding expression for a capped array is (Willis et al 1991)

$$E_{DSc} = \frac{\mu}{2\pi(1-\nu)} \left[2a_4 + a_1 \mathrm{Re} \ln\left\{ \frac{p(1-e^{-w})}{2\pi q} \right\} - a_3 \frac{4\pi h}{p} \mathrm{Re}\left\{ \frac{e^{-w}}{1-e^{-w}} e^{-2i\alpha} \right\} + a_5 \right]. \tag{6}$$

Here $w = \frac{2\pi h}{p} (1 - i\tan\theta)$, where θ is the angle which the line joining two dislocations in a dipole makes with the normal to the interface. The core cutoff radius q is assumed to be equal to b. We have calculated the total energies E_{Tu} and E_{Tc} of uncapped and capped layers respectively, containing the same density (=1/p) of dislocations. In Fig 1 we compare these

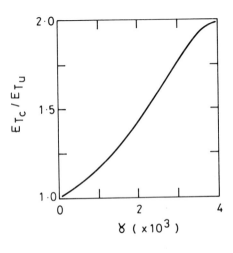

Fig. 1 Ratio E_{Tc}/E_{Tu} as a function of the strain relaxation $\gamma = -b_1/p$ for a 1000 Å $Ge_{0.1}Si_{0.9}$ layer

energies by plotting the ratio E_{Tc}/E_{Tu} as a function of the strain relaxation $\gamma = -b_1/p$ for 1000Å $Ge_{0.1}Si_{0.9}$ layers with $f_m = 0.0042$. The ratio is 1 for dislocation-free layers and increases monotonically with increase in strain relaxation. It approaches a value 2 for large values of strain relaxation.

3. CRITICAL LAYER THICKNESS AND EQUILIBRIUM STRAIN RELAXATION

According to equilibrium theory (Jain et al 1990), stable configurations are found by minimizing the total energy in the layer i.e. by finding triplets (f_{me}, h_e, p_e) which satisfy the condition

$$\frac{dE_T}{d(1/p)} = 0 \tag{7}$$

and the additional condition that the second derivative be positive. From Eqn (1), using Eqns (2), (3) and (4)

$$\frac{dE_T}{d(1/p)} = 2E_{DS} - 2p\frac{dE_{DS}}{dp} + 4\mu\frac{1+v}{1-v} b_1 h\left(f_m + \frac{b_1}{p}\right). \tag{8}$$

By differentiatiion of Eqn (6), we obtain

$$\frac{dE_{DSc}}{dp} = \frac{\mu}{2\pi(1-v)} \, Re\left\{\frac{a_1}{p}\left(1 - \frac{we^{-w}}{1-e^{-w}}\right) + a_3\frac{4\pi h}{p^2}\frac{e^{-w}(1-w) - e^{-2w}}{(1-e^{-w})^2}e^{-2i\alpha}\right\}. \tag{9}$$

If we now differentiate Eqn (9) again and solve for the second derivative of the total energy as $p\rightarrow\infty$, we obtain

$$Lim_{p\rightarrow\infty}\frac{d^2E_{Tc}}{d(1/p)^2} = \frac{2\mu h}{1-v}((1+2v+\cos 2\alpha)b_1^2 + (\cos 2\alpha - 1)b_2^2 + (v-1)b_3^2) \tag{10}$$

which has negative numerical value for $v = 0.3$. Therefore, contrary to the assertions of the theories of van der Merwe (see the discussion of this theory in Jain et al 1990) and Matthews and Blakeslee, the critical layer thickness is not that value of h for which first derivative of the total energy is zero as $p \rightarrow \infty$.

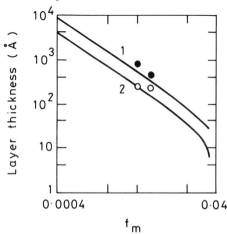

Figure. 2 Critical layer thickness of a capped layer(curve 1) and an uncapped layer (curve 2) as a function of f_m.

The true equilibrium critical layer thickness can be calculated by a method which will be described in another paper (Jain et al 1992). It is in fact smaller than that calculated using earlier theories but the difference is minor and can be neglected. This result is similar to that we obtained for an uncapped layer (Jain et al 1992). The calculated values of critical layer thickness for capped and uncapped layers are compared in Fig 2. The values for the capped layers are 2 to 4 times larger than the corresponding values for uncapped layers, depending on the value of the misfit parameter f_m. Experimental results of Houghton et al (1989) for capped layers are shown in the figure. The critical thickness lies between the open circles (pseudomorphic layers) and the closed circles (with dislocations, just above h_c) and agree well with our theory.

Equilibrium configurations (equilibrium values of h_e and b_1/p_e for a given f_m) are determined by finding values of p such that Eqn (7) is satisfied and the second derivative of the total energy is positive. We compare the values of $\gamma_e = -b_1/p_e$ for capped and uncapped layers, as a function of h, in Fig 3 for $Ge_{0.1}Si_{0.9}$ layers. Over the whole range of h shown in the figure, the extent of strain relaxation is considerably less in capped than uncapped layers.

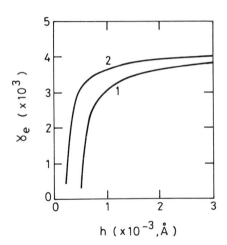

Figure. 3 Equilibrium strain relaxation γ_e as a function of layer thickness h for capped (curve 1) and uncapped (curve 2) $Ge_{0.1}Si_{0.9}$ layers.

4. PLASTIC FLOW IN METASTABLE LAYERS

For fixed layer thickness, ε_{re} is constant with time and the strain relaxation by plastic flow can be calculated by using the following relation (Jain et al 1992):

$$t = \frac{1}{C\mu^2(f_m - \varepsilon_{re} - \gamma_o)^2} \left\{ \ln\left(\frac{\gamma_o + \gamma}{f_m - \varepsilon_{re} - \gamma}\right) + \left(\frac{f_m - \varepsilon_{re} + \gamma_o}{f_m - \varepsilon_{re} - \gamma}\right) - \ln\left(\frac{\gamma_o}{f_m - \varepsilon_{re}}\right) - \left(\frac{f_m - \varepsilon_{re} + \gamma_o}{f_m - \varepsilon_{re}}\right) \right\}. \quad (11)$$

The difference between the rates of strain relaxation in capped and uncapped layers is demonstrated in Fig 4 for a 1000Å $Ge_{0.1}Si_{0.9}$ layer. Strain relaxation is predicted to occur at a considerably faster rate in capped layers.

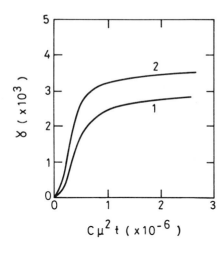

Fig. 4 Strain relaxation γ plotted as a function of reduced time, $C\mu^2 t$, for capped (curve 1) and uncapped (curve 2) 1000 Å $Ge_{0.1}Si_{0.9}$ layer.

To summarize, we have presented theoretical predictions of the critical thickness and strain relaxation of capped layers and shown that capped layers are much more stable than uncapped layers. There is good agreement between theoretical and experimental results.

APPENDIX: LIST OF SYMBOLS

In the text the subscript e added to any of the following parameters refers to the value of that parameter in an equilibrium configuration.

b	Burgers vector
b	Magnitude of Burgers vector
C	Phenomenological constant for plastic flow theory
E_D	Energy per unit area due to the fluctuating part of the strain field of an orthogonal periodic array of dislocations
E_{DS}	Energy per unit length of a dislocation line due to the fluctuating part of the field in a periodic array of dislocations
E_H	Homogeneous strain energy per unit area
E_T	Total elastic energy per unit area
f_m	Fractional lattice misfit parameter
h	Thickness of epilayer
h_c	Critical layer thickness
p	Dislocation spacing
q	Dislocation core cutoff radius
α	Angle between dislocation slip plane and normal to interface
θ	Angle between normal to interface and line joining the two dislocations in a dipole
ε_r	Relaxed misfit strain
γ	Strain relaxation
γ_0	Strain relaxation due to dislocation source density
μ	Shear modulus of elasticity
ν	Poisson's ratio

Conglomerate constants used in the body of the paper are:

$$a_0 = (b_1^2+b_2^2)\left\{\sin^2(\theta-\alpha) - \frac{1-2v}{4(1-v)}\right\}, \qquad a_1 = b_1^2+b_2^2+(1-v)b_3^2,$$

$$a_2 = b_1^2-b_2^2, \qquad a_3 = \frac{1}{2}(b_1^2+b_2^2),$$

$$a_4 = \frac{1}{2}(b_1^2+b_2^2)\left\{\sin^2(\theta-\alpha) - \frac{1-2v}{4(1-v)}\right\}, \qquad a_5 = (b_1^2+b_2^2)\{\cos^2\alpha - \sin^2(\theta-\alpha)\}.$$

REFERENCES

Garone P M, Venkataraman V and Sturm J C 1990 *IEDM 90, Tech. Dig.* 383

Gosling T J, Jain S C, Atkinson A, Willis J R and Bullough R 1992 to be published

Houghton D C 1990a *Appl. Phys. Lett.* **56** 460

Houghton D C 1990b *Appl. Phys. Lett.* **57** 1434

Houghton D C, Gibbings C J, Tuppen C G, Lyons M H and Halliwell M A G 1989 *Thin Solid Films* **183** 171

Houghton D C, Perovic D D, Baribeau J M and Weatherly G C 1990 *J. Appl. Phys.* **67** 1850

Jain S C and Hayes W 1991 *Semicond. Sci. Technol.* **6** 547

Jain S C, Gosling T J, Willis J R, Totterdell D H J and Bullough R 1992 *Phil. Mag. A.* to be published

Jain S C, Willis J R and Bullough R 1990 *Advances in Phys.* **39** 127

Kesan V P, May P G, Bassous E and Iyer S S 1990 *IEDM 90, Tech. Dig.* 637

Matthews J W and Blakeslee A E 1974 *J. Cryst. Growth* **27** 118

Patton G L, Comfort J H, Meyerson B S, Crabble E F, Scilla G J, de Fresart E, Stork J M C, Sun J Y C, Harame D L and Burghartz J N 1990 *IEEE Electron Device Letters* **11** 171

Willis J R, Jain S C and Bullough R 1990 *Phil. Mag. A* **62** 115

Willis J R, Jain S C and Bullough R 1991 *Phil. Mag. A* to appear in October

Defects created by electronic excitation in metals

Y. Quéré

L.S.I., École Polytechnique, 91128 Palaiseau, France

ABSTRACT : Contrary to what was previously believed, defects, and even tracks, can be created in metals by ion-irradiation-induced electronic excitation. For this to occur, the electronic stopping power must exceed a critical value (e.g. $\simeq$ 4 keV.Å^{-1} in Fe).

1. INTRODUCTION

When an ion slows down in a solid, it is classical to distinguish between its <u>nuclear stopping</u> NS (stopping power - $(dE/dx)_n = S_n$) and its <u>electronic excitation stopping</u> EES (stopping power S_e). Both contribute to defect formation in insulators where, in particular, EES is responsible for the creation of tracks as described by Price and Walker (1962) and studied by Bullough and Gilman (1966).

In pure metals, NS creates all kinds of defects whereas, due to the extremely rapid shielding of charges by free electrons, it has always been believed that EES was unable to create defects. This does not mean that EES was ignored in the defect formation process : the burst of EES energy ("thermal spike" : Seitz and Koehler 1956 ; "fission spike" : Quéré and Nakache 1959 ; Quéré 1963), was considered as a potential source of defect <u>recombination</u> ; of alloy <u>disordering</u> (Bloch 1960) ; and even of <u>amorphization</u> (Bloch 1962 ; Lesueur 1975 ; Audouard et al 1990) ; or of defect <u>clustering</u> (Weinberg and Quéré 1987). But again, in all these cases, no "direct" formation of crystal defects via EES was expected, nor reported, at least in pure metals. A review is presented here of experiments which show that, above a certain threshold in the value of S_e, this simple belief has to be revised.

2. HIGH EES IRRADIATIONS OF METALS

Dunlop et al (1989-1991) have irradiated various metals, or alloys, by various ions of different (high) energies both in Caen (at the CIRIL laboratory of GANIL) and in Darmstadt (GSI). By using high energy heavy ions (e.g. Xe or U ions of 3-5 GeV), values of S_e as high as $\simeq$ 8 keV.Å^{-1} can be produced in metals. In these conditions, it is clear that the EES is completely dominant over NS ($S_e/S_n \simeq 2000$).

In-situ low temperature measurements of the electrical resistivity increase $\Delta\rho$ have made it possible to determine in each case the <u>initial</u> increase $\Delta\dot\rho$ of damage per incident ion. It is obvious that in such measurements, the exact nature and configuration of the damage cannot be specified but it seems reasonable to compare the different values of $\Delta\dot\rho$ for a given metal irradiated by different ions of different S_e's.

In each of these irradiations, defects are created by nuclear collisions, with a contribution $\Delta\dot\rho_n$ to $\Delta\dot\rho$. For the ion and the metal considered, the corresponding displacement cross sections may be calculated with standard procedures (see Dunlop et al 1989a). Using commonly accepted values of defect resistivity in the metal considered, a theoretical determination of $\Delta\dot\rho_n$ is then deduced, and it is observed, without surprise, that at low values of S_e there is a good coincidence between experiment (i.e. $\Delta\dot\rho$) and calculation (i.e. $\Delta\dot\rho_n$). In these conditions it is expected — and observed — that, if both the measurement and the calculation are properly done, the ratio $\xi = \Delta\dot\rho/\Delta\dot\rho_n$ should be equal, or close to 1.

It is interesting now to study how this ratio ξ varies when larger values of S_e are used. Two cases may immediately be separated :

    i/ metals (Cu ; Iwase et al 1987) or alloys (Cu$_3$Au ; Dunlop et al 1990) in which ξ remains essentially constant (and equal to $\simeq$ 1) even for values of S_e as high as 8 keV.Å^{-1}, which implies that nuclear collisions are the only — or dominant — source of defect creation ;

    ii/ metals in which ξ is radically altered when large values of S_e are used.

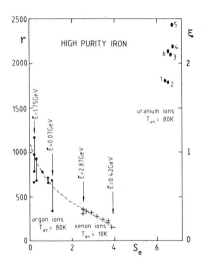

Figure 1 - Values of the quantity r (in µΩm) versus the electronic stopping power S_e (in keV/Å) in ion irradiated iron. The ratio r, equal to $\Delta\dot\rho/\dot c$ (where c is the calculated concentration of defects created in nuclear collisions by ions of fluence φ ; and $\dot c = dc/d\varphi$) is proportional to the ratio ξ (see text). Numerical values of ξ are plotted in the case where the resistivity of "defects" (e.g. Frenkel pairs) is taken as 1000 µΩm/unit concentration.
    Dunlop et al 1989b

A prototype of this second case is iron. As can be seen on figure 1, at <u>moderate electronic stopping power</u> a decrease of ξ is observed. This should be linked to the fact that subthreshold recombinations are remarkably easy in iron (Dural et al 1977) and also, more directly, to the fact observed by Dunlop et al (1989b) that, in a previously defect-doped iron, the defect concentration decreases under moderate S_e-value irradiation (Fig. 2). For <u>high electronic stopping power</u>, a definite enhancement of ξ takes place (Dunlop et al 1989b). This enhancement to even higher values of ξ (up to ≃ 12) can also be observed in metals such as Ti, Co and Zr (Dunlop et al 1991). Ga has also been reported to show this enhancement (Paumier et al 1989). The ξ behaviour of a number of metals is shown on figure 3.

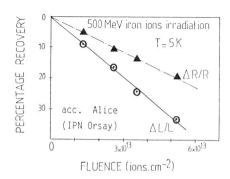

Figure 2 - If defects have been previously created, at 5K, in iron by a "low energy" Fe-ion-irradiation, which gives rise to an increase of both resistivity and length (i.e. volume), these two parameters *decrease* if the irradiation is continued, without heating, with Fe-ions of higher energy (i.e. higher electronic stopping power : here 1.3 keV/Å)

Dunlop et al 1989c

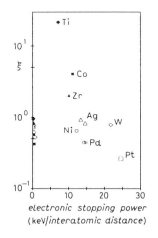

Figure 3 - A number of values of ξ ratio (see text) which describes production of defects by electronic excitation in various metals at various intensities of the electronic stopping power.
Dunlop et al 1991

The next question is that of the nature and morphology of the defects created at high values of S_e by the electronic excitation. This question is not yet answered in detail but transmission electron microscopy clearly reveals at least the presence of <u>tracks</u> (Barbu et al 1991) in high ξ-value metals.

These tracks may present two distinct morphologies. In an alloy such
as NiZr$_2$, they clearly consist of cylinders of <u>amorphized matter</u> which are
continuous along the path of the ions as soon as the value of S$_e$ is high
enough (Barbu et al 1991). In a pure metal like Ti after low temperature
irradiation and room temperature observation, they consist of a dense
alignment of small <u>dislocation loops</u> along the path (figure 4). These
loops, of Burgers vector 1/3 ⟨11$\overline{2}$0⟩, or 1/2 ⟨10$\overline{1}$0⟩, lie mostly in prismatic
{10$\overline{1}$0} planes (Henry et al 1992). For small values of the fluence, there is
a one-to-one correspondence between number of tracks and number of ions.
For higher values of the fluence, the number of tracks tends to a satura-
tion which indicates that fast ions, if able to create defects via EES
along their path, are also able to produce an annealing at a larger distance (Henry et al 1992).

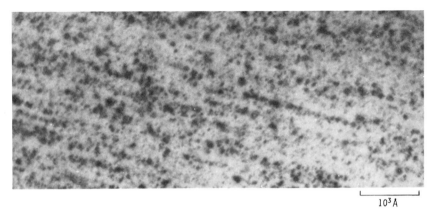

10^3 Å

Figure 4 - Transmission electron micrograph (bright field) of a titanium sample irradiated at
15 K by Pb ions of energy 4.4 GeV (electronic stopping power : 7 keV/atomic distance).
Alignments of defects (dislocation loops), parallel to the beam direction, are clearly visible.
      Henry et al 1992

What does distinguish those metals which are sensitive to electronic
excitation (like Ti, Zr, Fe) from those which are not (like Cu, Ag, W ; see
figure 3) ? The former have the common trait of exhibiting phase changes
and more generally displacive transformations (Legrand et al 1992). Let us
consider, in the frame of the Coulomb explosion model, positive ions created instantaneously after the passage of an ion. On a line of atoms (mass :
M) taken, for the sake of simplicity, as a one-dimensional chain (equidis-
tance : a), the ℓth atom receives an impulse Mv$_\ell$ from a force f$_\ell$(t) which
has to be taken, in a metal, as vanishingly short : f$_\ell$(t) = Mv$_\ell$(t) δ(t)
where δ is the Dirac distribution. This specific expression for the force
f$_\ell$(t) makes it possible to calculate the displacement u$_\ell$ as

$$u_\ell(t) = \sum_q \frac{\sin \Omega(q)t}{M \, \Omega(q)} F(q) e^{-iq\ell a} \qquad \text{(Dunlop et al 1991)}$$

where $\Omega(q)$ is the phonon spectrum and $F(q,\omega)$ the Fourier transform of $f_{\ell}(t)$. It is visible here that the presence of soft modes, i.e. of regions in the spectrum where Ω is small, favours large values of the displacement u_{ℓ}, i.e. potentially favourable conditions for point defect creation.

3. CONCLUSION

A short review has been presented of various experiments which aimed to study the effects of electronic excitation by fast heavy ions upon metallic targets in the regime of high stopping power S_e.

In some cases there is no observable effect (as previously believed) at least for the values of S_e which could be reached. In other cases, various - and possibly antinomic - effects have been described, according to the level of S_e as compared to a critical value S_{ec}. When $S_e < S_{ec}$, defects may be <u>annealed-out</u>. When $S_e > S_{ec}$, they may be both <u>created</u> (along the path of the ion) and <u>annealed-out</u> (further from the path), the defect creation being here far larger than that due to nuclear collisions.

A more complete description of these effects has recently been given by Dunlop and Lesueur (1991).

REFERENCES

Audouard A, Balanzat E, Bouffard S, Jousset JC, Chamberod A, Dunlop A, Lesueur D, Fuchs G, Spohr R, Vetter J and Thomé L 1990 <u>Phys. Rev. Lett.</u> <u>65</u> 875.

Barbu A, Dunlop A, Lesueur D and Averback RS 1991 <u>Europhys. Lett.</u> <u>15</u> 37

Bloch J 1960 <u>J. Nucl. Mater.</u> <u>1</u> 90

Bloch J 1962 <u>J. Nucl. Mater.</u> <u>6</u> 203

Bullough R and Gilman JJ 1966 <u>Phys. Rev.</u> <u>37</u> 2283

Dunlop A, Lesueur D and Dural J, 1989a <u>Nucl. Instr. Meth.</u> <u>B42</u>, 182

Dunlop A, Lesueur D, Morillo J, Dural J, Spohr R, and Vetter J 1989b <u>Comptes Rendus</u> <u>309</u> 1277

Dunlop A, Lesueur D, Jaskierowicz G and Schildknecht J 1989c <u>Nucl. Inst. Meth.</u> <u>B36</u> 412

Dunlop A, Lesueur D, Morillo J, Dural J, Spohr R and Vetter J 1990 Nucl. Inst. Meth. B48 419

Dunlop A, Lesueur D 1991 Proc. Int. Conf. Phys. Irrad. Metals, Siofok

Dunlop A, Legrand Ph, Lesueur D, Lorenzelli N, Morillo J, Barbu A and Bouffard S 1991 Europhys. Lett. 15 765

Dural J, Ardonceau J and Jousset JC 1977 J. Physique 38 1007

Henry J, Barbu A, Léridon B, Lesueur D and Dunlop A 1992 Proc. ICACIS-91 NIM

Iwase A, Sasaki S, Iwata T and Nihira T 1987 Phys. Rev. Lett. 58 2450

Legrand Ph. Dunlop A, Lesueur D, Lorenzelli N, Morillo J and Bouffard S 1992 Proc. ICACIS-91 NIM

Lesueur D 1975 Radiation Eff. 24 101

Paumier E, Toulemonde M, Dural J, Rullier-Albenque F, Girard JP and Bagdanski P 1989 Europhys. Lett. 10 555

Price PB and Walker RM 1962 J. Appl. Phys. 33 3400

Quéré Y and Nakache F 1959 J. Nucl. Mater. 1 203

Quéré 1963 J. Nucl. Mater. 9 290

Seitz F and Koehler JS 1956 Solid State Phys. 2 305

Toulemonde M and Paumier E 1992 To be published

Weinberg C and Quéré Y 1987 Materials Sc. Forum 15 943

Anisotropic diffusion of point defects: effects on irradiation induced deformation and microstructure

C.H. Woo, R.A. Holt[*] and M. Griffiths[*]

AECL Research, Whiteshell Laboratories, Pinawa, MB, Canada R0E 1L0
AECL Research, Chalk River Laboratories, Chalk River, ON, Canada K0J 1J0[*]

ABSTRACT

Mathematical equations are presented to calculate the influence of anisotropic diffusion on the irradiation response of metals. The equations reveal a very potent driving force for radiation damage evolution - Diffusional Anisotropy Difference (DAD). Intrinsic DAD is shown to explain a number of peculiar features of radiation damage evolution and irradiation growth of zirconium alloys. Extrinsic DAD (i.e., induced by stress) may play an important role in the irradiation creep of cubic metals.

1. INTRODUCTION

In this paper, effects caused by the Diffusional Anisotropy Difference (DAD) between the vacancies and SIAs are discussed. The scope is limited to effects related to the freely migrating point defects, leaving those related to point-defect clusters to other discussions (Woo and Singh 1991). Through DAD, (Woo 1983, 1985) fundamental point-defect properties can be related to the technologically important phenomena of irradiation creep and growth and the associated microstructural evolution. In the following, we first give a brief review of the theory, the predictions of which are then compared with experimental observations.

2. RATE THEORY MODEL

In the rate-theory model of radiation damage, reaction rates can be described by the theory of diffusion-influenced chemical reaction kinetics. Many authors, including Smoluchowski (1917), Onsager (1938) and Debye (1942) contributed to the development of this theory. Goesele and Seeger (1976) generalized the theory to include the case of bimolecular reactions between anisotropic diffusing species. Based on this theory, the conventional rate theory model has been generalized to include the effects of anisotropic diffusion (Woo 1987). It was found that the effect of anisotropic diffusion are very significant. The following is a brief account of the theory.

The reaction among the point defects and their annihilation at sinks, is assumed to be represented by the bimolecular reaction between two reacting species A and B, where at least one of the species is mobile by diffusion and the reaction product, C, has no influence on the further reaction process. Suppose the reactants are continuously being produced and annihilated, and have anisotropic diffusion tensors D^A and D^B, and an interaction potential E(r) between A and B. Goesele (1978) showed that under steady-state conditions, the rate of change in the spatially averaged concentrations C_A and C_B (in atomic fractions), can be approximated by

$$\dot{C}_A = \dot{C}_B = - \bar{D}\alpha C_A C_B \tag{1}$$

where α is the reaction constant, the dots over C_A and C_B denote the time derivative and $\bar{D} = (D_x D_y D_z)^{1/3}$ with D_x, D_y, D_z being the principal values of the relative diffusion tensor $\mathbf{D}$ $(= D^A + D^B)$, which is assumed to be constant in space and time. If B is an indestructible sink, then αC_B is the usual sink strength k_B^2 within the effective medium approximation (Goesele 1978). The lifetime τ_B of A from creation to annihilation at B is related to the corresponding reaction constants by

$$k_B^2 = \alpha C_B = (\bar{D}\tau_B)^{-1} \tag{2}$$

and the effective lifetime τ_{eff} in the effective medium is given by

$$\tau_{eff}^{-1} = \sum_B \tau_B^{-1} . \tag{3}$$

Expressions for typical sink geometries during irradiation have been derived and will be given in the following section.

3. REACTION CONSTANTS FOR VARIOUS SINK TYPES IN AN ANISOTROPIC DIFFUSIVE MEDIUM

The reaction constants of the following sinks are considered: the void, the straight dislocation, the dislocation loop and the surface and grain boundaries. To be specific, the diffusion tensor is assumed to have transverse isotropy, i.e., $D_x = D_y = D_a$, and $D_z = D_c$. To focus on the DAD effect, we subsume the bias effect due to elastic interactions in an effective sink of radius R (Debye 1942). We also define the parameter $p = (D_c/D_a)^{1/6}$ to measure the diffusional anisotropy.

For spheroidal voids with radii R_a $(=R_x = R_y)$ and R_c $(=R_z)$, the reaction constant for steady-state conditions can be derived from the results of Goesele and Seeger (1976)

$$\alpha = \frac{4\pi |p^{-4}R_c^2 - p^2 R_a^2|^{\frac{1}{2}}}{cs^1 (p^{-3}R_c/R_a)} + \frac{1}{2\sqrt{D\tau}} \left\{ p^2 R_a^2 + \frac{p^{-3}R_c R_a}{|p^{-4}R_c^2 - p^2 R_a^2|^{\frac{1}{2}}} cs^{-1}(p^3 R_a/R_c) \right\} \quad (4)$$

with

$$cs^{-1}(x) = \cos^{-1}(x) \quad |x| \le 1 \quad ; \quad cs^{-1}(x) = \cosh^{-1}(x) \quad |x| > 1$$

Equation (3) reduces for spherical voids to the earlier results (Woo 1988). Furthermore, for spherical void and isotropic diffusion, these results reduce to the expressions given in the conventional rate theory.

For straight dislocations, let λ be the angle made by the dislocation line with the c-axis of the diffusion field. The reaction constant per unit length of a cylindrical sink is given by (Woo 1988)

$$\alpha = p^{-2}(\cos^2\lambda + p^6 \sin^2\lambda)^{\frac{1}{2}}\alpha_o \quad (5)$$

where α_o is the corresponding reaction constant for isotropic diffusion. That α is a generalization of the isotropic diffusion results can be easily seen by putting $p = 1$.

The reaction constant of a large enough dislocation loop may be approximated by averaging the line direction of straight dislocations. Approximating an elliptical loop of radii R_a and R_c by a corresponding rectangle, the reaction constant of a loop with the loop-plane normal making an angle ϕ with the <c>-axis is given by

$$\alpha \approx \frac{p^3 + R_c/R_a(\sin^2\phi + p^6\cos^2\phi)^{\frac{1}{2}}}{p^2(1 + R_c/R_a)} \alpha_o^\ell \quad (6)$$

where α_o^ℓ is the reaction constant of the loop in an isotropic medium. It is easily seen that α is a function of the loop ellipticity, and that it is a generalization of the isotropic diffusion case where $p = 1$. It can also be seen that the reaction constants are different for different parts of the loop because of the orientation.

For surfaces, let us consider a pair of parallel planes separated by a distance 2d by a diffusive medium in which the point-defect lifetime is τ_1. Let ϕ be the angle made by the surface normal with the <c>-axis. The sink strength of the pair of surfaces is given by (Woo 1988)

$$k_s^2 = \frac{k_I}{d'} \left[\coth(k_2 d') - \frac{1}{k_I d'} \right]^{-1} \quad (7)$$

where

$$d' = \frac{pd}{(\sin^{-2}\lambda + p^6\cos^2\lambda)^{\frac{1}{2}}} \quad , \tag{8}$$

and k_i^2 is the sink strength of the effective medium enclosed by the surface pair, and is related to τ_i according to Eqn. (2). The corresponding reaction constant α is obtained by multiplying k_s^2 by d. These results reduce in the limit of isotropic diffusion to the conventional effective medium sink strengths for surfaces (Bullough, Hayns and Wood 1980). The foregoing discussion shows that different parts of the surface of a void have different reaction constants due to anisotropic diffusion. Equation (7) can be used to calculate the sink strengths of grain boundaries.

4. THE DAD EFFECT

The sink strengths and reaction constants given in Eqns. (4-7) are in general strong functions of p, the diffusional anisotropy parameter. It follows then that a DAD between the vacancy and the interstitial causes a difference in the reaction constants, and hence a bias in the point-defect absorption rate. The bias can be intrinsic due to the anisotropic crystal structure or extrinsic due to the perturbation on the crystal symmetry caused by an external deviatoric stress. DAD has several important consequences which cannot be derived from the conventional rate theory.

1. Voids are not neutral sinks, but have biases depending on void shape and diffusional anisotropy.
2. Edge dislocations and loops may not be biased towards interstitials. The bias of a dislocation depends on the line orientation and in the case of a loop its habit plane and ellipticity.
3. Surfaces and grain boundaries are also strongly biased sinks, depending on their orientation.
4. Different parts of the surface of a void, or of the line of a dislocation loop, may have different biases. Under steady-state irradiation damage condition, they may attain a shape different from the most thermodynamically favoured one and the shape may evolve size.

5. EXPERIMENTAL SUPPORT FOR THE DAD EFFECT

The experimental support for the DAD effect comes principally from the irradiation response of zirconium alloys. Irradiation growth of zirconium alloys exhibits complex characteristics in which large variations occur due to fairly small changes in microstructure. Typical irradiation growth behaviour, in material containing a high density of sinks in the grain interiors (relative to the grain size) reflects an expansion in the basal plane of the hcp structure and a contraction along the c axis (Fidleris 1987). This behaviour reflects the partitioning of SIA's to crystal defects (dislocations or loops) with an <a> Burgers vector, and vacancies to defects with a <c> components Burgers vector. Similar behaviour occurs whether the <c> component defects are network dislocations ($b = 1/3<11\bar{2}3>$) or faulted basal plane vacancy loops ($b = 1/6<20\bar{2}3>$).

Two peculiarities of the irradiation behaviour of zirconium are:
- the population of <a> loops formed during irradiation is of mixed character, i.e., vacancy and interstitial loops are both present (Griffiths 1988), and
- the growth behaviour is very complicated when the grain structure is very fine.

The latter is illustrated by the growth the Zr-2.5Nb alloy used for pressure tubes in CANDU reactors at 550K (Holt and Fleck 1990). The alloy is used in the extruded and cold-worked condition with a final stress relief. The grains are highly elongated and flattened in the radial direction. The standard pressure tube material exhibits the "normal" growth behaviour described above, with an expansion in the longitudinal direction - containing a very small proportion of <c> axes - and a contraction in the transverse direction - containing a high proportion of <c> axes, Figure 1A.

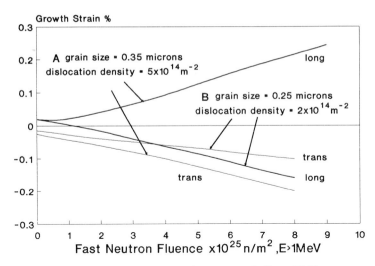

Fig. 1. Irradiation growth of Zr-2.5 wt% Nb alloy at 550 K

A small reduction in grain size and residual network dislocation density from cold-work drastically changes the anisotropy of growth, with negative strains in both the transverse and longitudinal directions, Figure 1B, in spite of almost identical crystallographic texture.

These observations cannot be explained by the usual dislocation bias arising from the size interaction of SIA's with dislocation strain fields and strongly suggest the role of DAD. The normal growth behaviour, combined with the fact that the <c> component defects usually lie close to the basal plane, while <a> type defects lie close to the <c> direction, suggest that $p_v > p_i$.

A model for irradiation growth employing the relationships given above, and fitting the ratio p_v/p_i to growth data for coarse grained Zircaloy-2 (Rogerson 1988) gives a semi-quantitative explanation for the observed growth behaviour (Holt 1988). Figure 2 shows that there is a calculated regime of grain thickness and network dislocation density in which both transverse and longitudinal growth are negative. The calculations show that <a> loops have little preference for either vacancies or SIA's because their line direction is partly in the basal plane and partly close to the <c> direction.

One consequence of these results is the prediction that voids in zirconium should experience a net vacancy flux at surfaces close to the basal plane and a net interstitial flux at surfaces containing the <c> axis. This coincides with recent HVEM studies of void growth in zirconium at 570K (Griffiths et al. 1991). Typically, the voids are observed to nucleate as thin discs on the basal planes which subsequently increase in thickness and decrease in width, Figure 3.

6.   DISCUSSION - IMPLICATIONS OF DAD FOR CUBIC MATERIALS

The theory of irradiation induced dimensional changes is undergoing a radical review with recent suggestions that the efficiency of displacement damage cascades at producing free point defects is very much lower than previously believed (Muller et al. 1988) - of the order of a few percent instead of close to one. The implication is that much larger driving forces are required (segregating vacancies and interstitials) to explain observed dimensional changes produced by swelling and irradiation creep than was previously believed.

For example, irradiation creep due to SIPA, has previously been ascribed to the second order SIA/dislocation elastic interaction. If free point defect generation rates are an order of magnitude or more less than traditionally calculated displacement rates, this effect is far too small to account for observed creep rates. SIPA induced by elasto-diffusion via DAD is, however (potentially) a much stronger effect and has recently been shown to account for both observed magnitude of irradiation creep in austenitic steels (Woo et al. 1991).

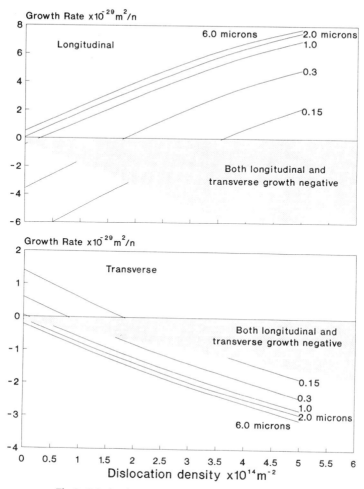

Fig. 2. Calculated growth of Zr-2.5 wt% Nb alloy at 550 K as a function of grain size and dislocation density.

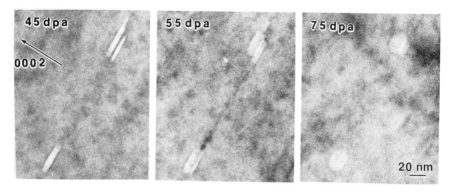

Fig. 3. Anisotropic void growth in Zirconium during electron irradiation in a HVEM at 570 K

7. CONCLUSIONS

Equations are presented to calculate the influence of anisotropic diffusion on the irradiation response of metals. The equations reveal a very potent driving force for radiation damage evolution - diffusional anisotropy difference (DAD). Intrinsic DAD is shown to explain a number of peculiar features of radiation damage evolution and irradiation growth of zirconium alloys. Extrinsic DAD (i.e., induced by stress) may play and important role in the irradiation creep of cubic metals.

ACKNOWLEDGEMENT

This work was financially supported by the CANDU Owners' Group (COG).

REFERENCES

Bullough R, Haynes M R and Wood M H 1980 *J. Nucl. Mat.* 90 44.
Debye P 1942 *Trans. Electrochem. Soc.* 82 265.
Fidleris V, Tucker R P and Adamson R B 1987 *7th International Conference on Zirconium in the Nuclear Industry* (ASTM Philadelphia) eds. R B Adamson and   L F P Van Swam STP 939 49.
Griffiths M, Woo C H, Frank W, Styles R C and Phillipp F 1991 to be published.
Gosele U, 1978 *J. Nucl. Mater.* 78 83.
Gosele U, and Seeger A 1976 *Phil. Mag.* 34 177.
Holt R A 1988 *J. Nucl. Mater.* 159 310.
Holt R A and Fleck R G 1988 *8th International Conference on Zirconium in the Nuclear Industry* (ASTM Philadelphia) eds. L F P Van Swam and C M Eucken STP 1023 705.
Muller A, Naundorf V and Macht M-P 1988 *J. Appl. Phys.* 65 3445 Onsager 1938 Phys Rev. 54 554.
Onsager L 1338 *Phys. Rev.* 54 554.
Rogerson A 1988 *J. Nucl. Mater.* 159 43.
Smoluchowski M V 1917 *Z Phys. Chem.* 92 219.
Woo C H and Gosele U 1983 *J. Nucl. Mat.* 119 219.
Woo C H 1985 *J. Nucl. Mater.* 131 105.
Woo C H 1987 *Radiation Induced Changes in Microstructure* ed. F.A. Garner et al. (ASTM Philadelphia) STP 955 70.
Woo C H 1988 *J. Nucl. Mat.* 159 237.
Woo C H and Singh B 1991 *Phil. Mag.* in press.
Woo C H, Singh B and Garner F A 1991 to be published.

Point defects, elastic moduli and melting in metals

N. H. March
Theoretical Chemistry Department, University of Oxford, 5 South Parks Road,
Oxford, OX1 3UB, England.

ABSTRACT: Recent progress in the theory of point defect energetics will be surveyed. In particular, a thermodynamic formula of the writer for the vacancy formation energy in a 'hot' close–packed crystal is compared and contrasted, for both s–p metals and transition metals, with a semi–empirical ground–state formula of Miedema. Though these two approaches give insight into observed trends, quantitative accuracy for a particular crystal is inevitably sacrificed. Therefore, fully quantitative numerical studies of the electron theory of defect energies in Al by de Vita and Gillan are summarized, both vacancy and self–interstitial energetics being considered by these workers. The interest in extending their work to the body–centred–cubic alkalis Na and K is stressed, with its possible relevance to diffusion mechanisms in open crystal structures.

1.   INTRODUCTION

It is, of course, a privilege to present this survey paper at this meeting honouring Dr Ron Bullough. Much of my own group's work on point defects has been carried out over more than two decades in close collaboration with AEA Harwell. However, Ron and I first met at Sheffield University, where Ron was working with Professor Bruce Bilby in the Metallurgy Department while I was based in Physics. His interests then were in dislocation theory (Bilby *et al.* 1955; Bullough and Bilby, 1956) while mine were already building up in the point defects area (Alfred and March, 1955). Later, our interests came together, notably though our mutual interaction with Dr Roy Perrin, and papers by our two groups resulted (Brown *et al.* 1971). Ron's major interests in the theory and technological relevance of radiation damage provided a lot of stimulation to our continuing studies on point defects.

In the present survey, some focus will be provided by a simple phenomenological 'model' for the estimating vacancy formation energy E_v in a hot close–packed crystal from experimental data on the liquid just above its melting temperature T_m; contact being established here with the semi–empirical approach of Miedema (1979). Then, full electron theory calculations made very recently on Al by de Vita and Gillan (1991) will be summarized; these workers not only calculated E_v but also the energy of vacancy migration E_m. Their preliminary results on the Al self–interstitial will also be referred to.

Brief reference will be finally made to the technological importance of the interaction between point defects, and to the continuing interest of diffusion mechanisms in crystalline solids: especially non–close–packed materials.

2. PHENOMENOLOGICAL 'MODEL' RELATING VACANCY FORMATION ENERGY IN HOT CLOSE–PACKED CRYSTALS TO THERMO–DYNAMIC PROPERTIES

Let us adopt the 'criterion' for melting of close–packed crystals, following March (1987), that the solid–liquid phase transition occurs when the internal energy (E_v) required to create a localized hole in the hot crystal becomes equal to the increase in internal energy needed to expand the liquid by one atomic volume (i.e. neglecting relaxation round the vacant site in the hot crystal). If U denotes the internal energy of the liquid with atomic volume Ω and N atoms then the above criterion becomes

$$E_v = U(N\Omega + \Omega) - U(N\Omega) = \Omega \left(\frac{\partial U}{\partial V} \right)_{T_m} \qquad (2.1)$$

with T_m the melting temperature, and $V = N\Omega$ is the total liquid volume.

Using the thermodynamic formula for the difference $c_p - c_v$ of the specific heats:

$$c_p - c_v = -\frac{T}{\rho V} \left(\frac{\partial p}{\partial T} \right)_V^2 \left(\frac{\partial V}{\partial p} \right)_T , \qquad (2.2)$$

with ρ the atomic number density it is a straightforward matter to express this in terms of pressure p and $(\partial U/\partial V)_T$ as

$$\frac{c_p - c_v}{S(0)} = \left\{ \frac{p}{\rho k_B T} + \frac{1}{\rho k_B T} \left[\frac{\partial U}{\partial V} \right]_T \right\}^2 k_B. \qquad (2.3)$$

Here $S(0)$ is the long–wavelength limit of the liquid structure factor $S(k)$, again related to thermodynamics by

$$S(0) = \rho k_B T K_T \qquad (2.4)$$

with K_T the isothermal compressibility. Evaluating eqns (2.3) and (2.4) at the melting temperature T_m and noting that there $p/\rho k_B T_m \ll 1$, one finds

$$\frac{1}{\rho k_B T_m} \left[\frac{\partial U}{\partial V} \right]_{T_m} = \left[\left\{ \frac{c_v(\gamma-1)}{k_B S(0)} \right\}_{T_m} \right]^{1/2} \qquad (2.5)$$

where γ is the ratio of specific heats, c_p/c_v. Using eqn (2.5) in eqn (2.1) yields at $T = T_m$ the desired expression for the vacancy formation internal energy E_v:

$$\frac{E_v}{k_B T_m} = \left[\left\{ \frac{(\gamma-1)c_v}{k_B S(0)} \right\}^{1/2} \right]_{T_m} . \qquad (2.6)$$

This formula[1] has been evaluated using empirical data for eight close–packed non transition metals (see also Table 3.1) by Rashid and March (1989). Their results may be summarized by quoting the average value of $E_v/k_B T_m$ from experiment of 9.4 ± 1.8 while the thermodynamic term (2.6) contributes an average of 8 ± 1.

3. CONNECTION WITH SEMI–EMPIRICAL TREATMENT OF MIEDEMA

Following Alonso and March (1989), let us first note that the simplest form (see Miedema and Boom, 1978; Miedema, 1979) of Miedema's treatment is to write

$$E_v = Q\Omega^{2/3}n_b \qquad (3.1)$$

where Ω is the atomic volume, n_b is the electron density at the boundary of the Wigner–Seitz cell in the pure metals and Q is an empirical constant.

---

[1] The very close relation between this formula (2.6) and conventional pair potential theory has been exhibited explicitly by the writer (March, 1987).

It is noteworthy, in contrast to eqn (2.6), that eqn (3.1) is essentially a 'ground–state' formula, involving the microscopic boundary electron density n_b. To make contact with eqn (2.6), let us next eliminate this microscopic quantity from eqn (3.1) using the empirical relation (Alonso and March, 1985; see also Alonso and Silbert, 1987) between the compressibility[2] and n_b at melting, namely

$$K_{T_m} = K_m = a(\Omega N_a n_b^2)^{-1} \tag{3.2}$$

where N_a is Avogadro's number and a an empirical constant. Substituting eqn (3.2) into eqn (3.1) then yields

$$E_v = Q \, \Omega^{1/6} \left[\frac{a}{N_a K_m} \right]^{1/2} \tag{3.3}$$

Again measuring E_v in units of $k_B T_m$ and utilizing eqn (2.4) to eliminate K_m, the result (3.3) may be rewritten as

$$\frac{E_v}{k_B T_m} = \frac{Q}{\Omega^{1/3}} \left[\frac{a}{N k_B T_m} \right]^{1/2} \frac{1}{\{S(0)\}^{1/2}_{T_m}}. \tag{3.4}$$

It is to be stressed (Alonso and March, 1989) that $\{S(0)\}^{\frac{1}{2}}_{T_m}$ now appears in both eqns (2.6) and (3.4). But it should also be noted that Miedema's formula (3.1) correlates only roughly with experiment for a single value of the 'constant' Q. A better correlation is obtained through the formula

$$E_v = Q \left[\frac{Z_{eff}}{Z} \right]^{3/5} \Omega^{2/3} \, n_b \tag{3.5}$$

where Q takes one value for non–transition elements and another for transition metals. In eqn (3.5), Z is the total number of valence (s+d) electrons and Z_{eff} is an effective valence (equal to the product $\Omega \, n_b$). The latter quantity has been discussed by a number of workers, a fairly recent treatment being that of Flores *et al.* (1981), with earlier references quoted there.

One obvious difference between the thermodynamic formula (2.6) and Miedema's treatment is that the latter does not need to distinguish between close–packed and open structures. Since eqn (2.6) is applicable only to close–packed crystals, one must of course restrict comparison to such materials. For the eight s–p metals

Table 3.1

Values of $E_v/k_B T_m$ calculated from the thermodynamic eqn (2.6) compared with experimental data and with Miedema's (1979) results. (After Alonso and March, 1989).

Metal	Experiment (from Miedema (1979))	Eqn (2.6)	Miedema (1979)
Cu	8.87	8.00	
Ag	9.26	8.43	
Au	8.10	8.93	
Mg	7.30	6.71	6.78
Zn	8.68	7.44	9.38
Cd	8.30	7.71	12.55
Al	8.12	6.76	7.73
Pb	9.61	8.03	8.81

[2]Using embedded atom theory which transcends pair potentials, Johnson (1988) has argued that E_v correlates more strongly with shear than bulk moduli (see also March (1989)).

already referred to, Table 3.1 compares formulae (2.6) and (3.1). A similar comparison has been made for the transition metals Co, Ni and Pd. The average prediction for $E_v/k_B T_m$ from eqn (2.6) is now 11.5 ± 0.6, while experiment is again 9.0 ± 0.7. Miedema's prediction is somewhat better for these metals (7.6 ± 0.4) but it does have semi–empirical quantities Q and Z_{eff} in it.

4. NUMERICAL CALCULATIONS VIA ELECTRON THEORY FOR POINT DEFECTS IN Al

Having attempted in sections 2 and 3 to expose an overall pattern in the behaviour of the ratio $E_v/k_B T_m$; with consequent sacrifice of accuracy for individual materials, it is of obvious interest to enquire what detailed electron theory predicts for a specific material. Below, as an example, we shall summarize the numerical study of de Vita and Gillan (1991) on Al, as the vacancy in this material represents one of the simplest examples of a defect in a technologically important material.

Gillan (1989) had previously obtained already useful accuracy for E_v from a full electronically self–consistent pseudopotential treatment. The aim of the later work of de Vita and Gillan was to re–examine the energetics of the vacancy problem using a fully ab initio non–local pseudopotential, non–locality being avoided in the earlier study of Gillan. In addition to the calculations of E_v and the migration energy E_m for the vacancy in Al, de Vita and Gillan also report preliminary results for the fully relaxed self–interstitial in Al.

4.1   Brief Technical Details

As in Gillan (1989), de Vita and Gillan performed their calculations in periodically repeated geometry, using up to 27 atomic sites in the unit cell, with a plane–wave basis set. Simultaneous relaxation of the electronic and ionic coordinates to the global energy minimum was performed using the conjugate gradients technique. Brillouin zone sampling was carried out by the scheme of Monkhorst and Pack (1976). The non–local pseudopotential already referred to was due to Bachelet *et al.* (1982), in the representation of Kleinman and Bylander (1982), with s and p non–locality. It should be added here that the conjugate gradients method described by Gillan (1989) has in common with the scheme of Car and Parrinello (1985) that minimization is carried out of the total energy functional with respect to all the plane wave coefficients, with the constraint that all the occupied orbitals remain orthonormal.

4.2   Results For Vacancy Energetics

The most reliable value obtained by de Vita and Gillan for E_v from their ground–state calculation is 0.55eV, to be compared with the experimental enthalpy of formation of 0.66eV. These workers also calculated the volume of formation Ω_f which they find to be 0.71 times the atomic volume Ω. Experimental values have been discussed by Seeger *et al.* (1971); they range from 0.55Ω to 0.68Ω, the latter value of Harrison and Wilkes (1971) having an estimated error of 0.1Ω. Thus there seems again good agreement between theory and experiment.

de Vita and Gillan have also studied by their procedure the migration energy of a vacancy in Al. Specifically they have calculated the ground–state energy of the relaxed system in which a migrating Al atom is fixed midway between two vacancies located in nearest neighbouring perfect lattice sites. They find a migration energy $E_m = 0.57$eV, to be compared with the experimental value of 0.62eV proposed by Seeger *et al.* (1971).

4.3   The Self–Interstitial in Al

de Vita and Gillan (1991) report also preliminary results for the self–interstitial in
Al. This was calculated in the 27 site supercell, for a single interstitial placed at the
octahedral site.   Technical difficulties arise from the strong long–range ionic
displacement field, but appear to have been surmounted. These workers obtain a
formation energy of the self–interstitial of 2.8eV, to be compared with the
experimental values of 3.2 ± 0.5eV given by Schilling (1978) as obtained from the
Frenkel defect formation energy of 3.9 ± 0.5eV, and the vacancy formation energy of
0.66eV. The same formation energy as obtained without relaxing the ions is found
by de Vita and Gillan to be ~10eV, which leads to a very large relaxation energy
~7eV. The calculated displacement of the six nearest neighbour atoms found by
these workers amounts to 20% of the distance from the self interstitial. However, de
Vita and Gillan qualify their findings by pointing out that both experiment· and
previous calculations indicate that the stable configuration of the self–interstitial is
not the octahedral configuration they have treated (Schilling, 1978; Sindzingre,
1988), though the energy differences between competing configurations are expected
to be small.

5.   DISCUSSION AND SUMMARY

Sections 2 and 3 have been concerned with understanding trends of vacancy
formation energies E_v through both s–p metals and transition metals.   Both
formulae (2.6) and (3.4) write the ratio $E_v/k_B T_m$, with T_m the melting temperature
in terms of $S(0)|_{T_m} = \rho \, k_B T_m K_{T_m}$, with K_{T_m} the isothermal compressibility.
However, this latter quantity does vary significantly on melting, and this must be
borne in mind in assessing the quantitative usefulness of these approaches. They do,
nonetheless, afford some degree of insight into the vacancy energy E_v, though one,
the thermodynamic formula (2.6), is a 'hot crystal' formula, while the Miedema eqn
(3.1) is a 'ground–state' result.   As de Vita and Gillan point out, experimental
values of E_v, E_m and Ω_f presently come from measurements at high temperatures,
whereas their own work, in common with Miedema's, is a ground–state study.
However, general arguments suggesting that the temperature dependence of
formation and migration enthalpies should be weak have been presented by Gillan
(1981) and Harding (1985, 1990).   As de Vita and Gillan note, the only direct
evidence on Al, for which their detailed electron theory summarized in section 4 has
been worked out, comes from molecular dynamics simulations of Jacucci *et al.*
(1981), which indicate that the enthalpy of formation of the vacancy might be
increased by about (1/10)eV in going from T = 0K to 860K.

We want to conclude by referring to the continuing interest for technological
problems of (a) point defect interactions among themselves (for mechanical
properties: see Greenwood, 1991) and (b) point defect–solute interactions (see the
review by March, 1978; see also the book by Allnatt and Lidiard, 1992). In (b), a
current example of some interest concerns impurities like P in bcc Fe: here the
studies of Alfred and March (1956) and of LeClaire (1962), when modified to include
a Thomas–Fermi screening length depending directly on electronic structure,
through the electronic density of states $N(E_f)$ at the Fermi energy E_f, may prove
relevant.   Another related area of interest is that of vacancy formation in alloys
and/or mixtures, recently reviewed briefly by the writer (March, 1990).

But one other direction in which further progress might come is prompted by the
quantitative electron theory study of de Vita and Gillan (1991) on the
closed–packed Al crystalline ground state.   Our own early work with Dr Ron
Bullough's group (Brown *et al.* 1971) on diffusion in body–centred–cubic Na has
been taken somewhat further subsequently by Flores and March (1982). In the

model of complete relaxation adopted in this latter work, the self–interstitial formation energy comes out only a few (1/100)ths of an eV higher than the vacancy energy E_v. Then, the observation of Brown *et al.* (1971) that the migration energy of the self interstitial in the bcc (Na) lattice is very small for geometrical (rather than force field) reasons opens up the possibility that diffusion in Na and the heavier alkalis might be due, at low temperatures, to predominantly an interstitial mechanism. Whether or not this turns out to be true, it would seem to be now of considerable interest to apply the method of de Vita and Gillan (1991) to both the vacancy and the self interstitial in the ground state of Na and K. This might contribute then to settling what has been a long–standing issue of diffusion mechanisms in bcc crystals.

ACKNOWLEDGEMENTS

I wish to thank Dr J. Matthews for inviting me to present this paper at the Bullough Symposium. I am greatly indebted to Dr A. Lidiard for continuing to interest my group in problems relating to technology in this general area over two decades. Finally I am grateful to Prof M. Gillan for showing me his work with Mr de Vita on Al in advance of publication and to Dr A. Harker for much support and encouragement during the course of the present work on interatomic forces and point defects.

REFERENCES

Alfred L C R and March N H 1955 *Phil.Mag.* **46** 759; 1957 *ibid* **2** 985; 1956 *Phys. Rev.* **103** 877
Allnatt A R and Lidiard A B *Atomic Transport in Solids* (Cambridge University Press) to appear 1992
Alonso J A and March N H 1985 *Surf.Sci.* **160** 509; 1989 *Phys.Chem.Liqs* **20** 235
Alonso J A and Silbert M 1987 *Phys.Chem.Liqs.* **17** 209
Bachelet G B, Hamann D R and Schlüter M 1982 *Phys.Rev.* **B26** 4199
Bilby B A, Bullough, R. and Smith E 1955 *Proc.Roy.Soc.* **A231** 263
Brown R C, Worster J, March N.H, Perrin R C and Bullough R 1971 *Z. Naturforsch* **26A** 77; 1971 *Phil.Mag.* **23** 555
Bullough R and Bilby B A 1956 *Proc.Phys.Soc.* **B69** 1276
Car R and Parrinello M 1985 *Phys.Rev.Lett.* **55** 2471
de Vita A and Gillan M J 1991 *private communication and to be published*
Flores F, Gabbay I and March N H 1981 *Chem.Phys.* **63** 391
Flores F and March N H 1982 *J.Phys.* **F12** L133; 1982 *Proc.Int.Conf.on Lattice defects in metals* Kyoto Nov. 1981
Gillan M J 1981 *Phil.Mag.* **A43** 301; 1989 *J.Phys.Cond.Matter* **1** 689
Greenwood G W 1991 *ICTP Trieste Lectures; see also these Proceedings*
Harding J H 1985 *Phys.Rev.* **B32** 6861; 1990 *Rep.Prog.Phys.* **53** 1403
Harrison E A and Wilkes P 1971 1st *Eur.Conf.Cond.Matter* (Geneva: EPS) p. 67
Jacucci G, Taylor R, Tenenbaum, A and van Doan N 1981 *J.Phys.F.* **11** 793
Johnson R A 1988 *Phys.Rev.* **B37** 3924
Kleinman L and Bylander D M 1982 *Phys.Rev.Lett.* **48** 1425
Le Claire A D 1962 *Phil.Mag.* **7** 141
March N H 1978 *J.Nucl.Mats.* **70** 490; 1987 *Phys.Chem.Liq.* **16** 209; *ibid* **17** 1; 1989 *Phys.Rev.* **B40** 3356; 1990 *J.Chem.Soc.Faraday Trans.* **86** 1203
Miedema A R 1979 *Zeit.Metallk.* **70** 345
Miedema A R and Boom R 1978 *Zeit.Metallk.* **69** 183
Monkhorst H J and Pack J D 1976 *Phys.Rev.* **B13** 5188
Rashid R I M A and March N H 1989 *Phys.Chem.Liq,* **19** 41
Schilling W 1978 *J.Nucl.Mater.* **69** 465
Seeger A, Wolf D and Mehrer H 1971 *Phys.Stat.Solidi(b)* **48** 481
Sindzingre P 1988 *Phil.Mag.* **57** 877

Point defect behaviour and its influence on some macroscopic properties

G W Greenwood

School of Materials, University of Sheffield, Mappin Street, Sheffield, S1 3JD, UK

ABSTRACT: Amongst the many ways in which point defects can influence macroscopic phenomena, two areas are considered here. One concerns the behaviour of vacancies and consequent dimensional changes when two elements inter-diffuse at different rates across their interface. The other relates to the vacancy source and sink distribution which leads to anisotropy of creep strength in materials where grain shape is not equiaxed. Both these areas illustrate the importance of a knowledge of vacancy sources and sinks.

1. INTRODUCTION

The creation, absorption and behaviour of vacancies is strongly temperature dependant and is influenced by other crystal defects, chemical factors, microstructure and external conditions such as environment, irradiation and superimposed stresses (Bueren 1960). Ron Bullough's work has touched on all these areas and it is a pleasure to acknowledge his stimulating contributions. He has analysed in depth several of the interactive effects on vacancy behaviour showing in an elegant way how changes in volume and in mechanical properties may ensue (Bullough and Finnis 1983).

The kinetics of many solid state processes can frequently be related to the role of vacancies though often a number of stages must be taken into account. These stages may be individually identified and the mode of coupling between them can be important. We see, for example, stress driven solute segregation and the significance of stress fields near crack tips (Rauh et al 1989, Rauh and Bullough 1985).

Vacancies may be created by quenching, irradiation, chemical potential gradients and deformation. In pure metals surfaces, voids and grain boundaries appear to operate freely as vacancy sources and sinks (Barnes et al 1958) in so far as they rarely introduce an interfacial step sufficient for dynamic control of any process but this situation can be substantially altered for alloys by solute segregation or dispersed particles (Harris et al 1969).

Extensive data exists on vacancy mobility through measurements of self and solute diffusion coefficients. Knowledge of the kinetics of movement however must be used in conjunction with the identification of diffusion paths to evaluate many rate processes. This is achieved by determining surfaces of chemical equi-potential to which the most probable diffusion paths are always orthogonal. Vacancy sink operation is then important. Thus the processes of vacancy emission, diffusion and absorption must often all be considered for the interpretation of macroscopic changes at elevated temperatures.

The example presented in Section 2 is directly related to the process of diffusion bonding of dissimilar materials where unequal rates of interdiffusion may cause dimensional changes and weakening through vacancy condensation to form voids.

The second example in Section 3 concerns the creation and absorption of vacancies at grain boundaries when small stresses act on materials at elevated temperatures (Burton 1977) and is related to the development of unidirectionally creep resistant alloys by the generation of elongated grains. When source and sink behaviour is inhibited then suggestions are made for a modified dependence of creep strength on grain size and shape.

2. VACANCY BEHAVIOUR DURING INTERDIFFUSION

Diffusion bonding by placing two materials together at elevated temperature sometimes forms a critical stage in manufacturing processes. Even in the simplest case however where two different metals are in contact their intrinsic diffusion coefficients may differ substantially and this condition is identified through the Kirkendall Effect (Smigelskas and Kirkendall 1947). The difference in atomic fluxes J_a and J_b of A and B atoms is made up by the flux of vacancies J_v so that the algebraic sum of the fluxes $J_a + J_b + J_v = 0$. Taking a sandwich of the metals A-B-A, then inert markers placed at the two interfaces each move a distance ΔX_m in time t at elevated temperature. The instantaneous velocity of marker movement $v = (d/dt)\Delta X_m$. Taking the intrinsic diffusion coefficients, which are concentration dependant, D_a and D_b of the A and B atoms respectively, the atom fluxes are then $J_a = -D_a(\partial c_a/\partial x)$ and $J_b = -D_b(\partial c_b/\partial x)$ where c_a and c_b are the atomic concentrations with $c_a + c_b = 1$. It follows (Beuren 1960) that $v = (D_a - D_b)(\partial c_a/\partial x)$ when c_a is the atomic concentration of A atoms in the plane of the markers.

In the application of the above formulae the equation of continuity also requires a rate of vacancy generation $(\partial c_v/\partial t) = (\partial/\partial x)[(D_a - D_b)(\partial c_a/\partial x)]$. The question immediately raised concerns the behaviour of these vacancies and the sinks to which they eventually diffuse. Many observations have revealed their tendency to condense to form voids (Le Claire and Barnes 1951, Buckle and Blin 1952, Balluffi 1954) but with some indication of the importance of microstructural features. Void formation appears to be decreased in finer grain material, though often grain growth is found to occur. The superimposition of a modest hydrostatic pressure of about 50 MPa has invariably been noted to suppress void formation entirely (Clay and Greenwood 1972) suggesting a critical nucleation radius for void formation of about 0.1 μm without such pressure. Voids form in progressively widening zones with their boundaries indistinctly defined but parallel to the original interface. This results in the volume change being manifest almost entirely as a length increase to which the original interface is perpendicular (Reed-Hill 1964). In the experimental technique it is essential to arrange the sandwich such that the voids are formed in the outer metal with the marker movement measured across the central void-free region.

If E is the fraction of vacancies collected into voids, the corresponding length increase perpendicular to each interface (Kheder 1976) is given by

$$\Delta L = \int_0^t \int_0^\infty E\,(\partial c_v/\partial t)\,dxdt$$

It suffices for the evaluation of E after any time t simply to compare marker movement with overall length change since $E = \Delta L/\Delta X_m$.

Some results (Kheder 1976) from the extensively investigated copper-nickel system are given in Figs 1 and 2. The marker movement ΔX_m, measured across the nickel rich zone in which porosity has not occurred, shows that over the temperature range 700 to 1033 °C ΔX_m is always proportional to the square root of time (Fig. 1). In contrast with this, Fig 2 illustrates that a comparable relationship $\Delta L \propto t^{1/2}$ only holds accurately at the lower temperatures and for limited times. Moreover, the extent of overall void formation decreases with increase in temperature of this range. Taking the ratio $\Delta L/\Delta X_m$ it is seen that whilst the $t^{1/2}$ relationship is followed for both ΔL and ΔX_m the efficiency of vacancy collection into voids is close to 100 per cent.

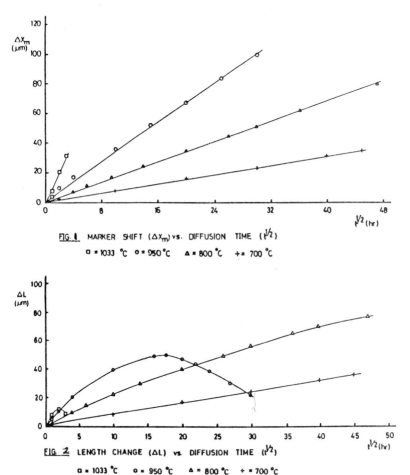

FIG. 1 MARKER SHIFT (ΔX_m) vs. DIFFUSION TIME ($t^{1/2}$)

□ = 1033 °C   o = 950 °C   ▲ = 800 °C   + = 700 °C

FIG. 2 LENGTH CHANGE (ΔL) vs DIFFUSION TIME ($t^{1/2}$)

□ = 1033 °C   o = 950 °C   ▲ = 800 °C   + = 700 °C

Metallographic observations show that the region of void formation widens as the diffusion zone extends. Voids tend to become larger but fewer in number as the chemical concentration gradient reduces. Then the rate of vacancy generation is insufficient to cause overall void growth and smaller voids progressively disappear.

The void coarsening process is amenable to analysis in the situation where vacancies condense exclusively into voids. The void spacing at any instant can be shown (Kheder 1976) to be of the order of $[10^2 \tilde{D}\gamma\Omega/(\partial c_v/\partial t)kT]^{1/3}$ where $\tilde{D}$ may be determined (Reed-Hill 1964) by Boltzmann-Matano analysis, Ω is the atomic volume, γ is the surface energy per unit area surrounding the voids and kT has the usual meaning. The rate of vacancy generation $(\partial c_v / \partial t)$ is time, temperature and position dependent.

As shown in Fig 2, the porosity attains a maximum value before steadily decreasing at sufficiently high temperatures and long times. The vacancies are then absorbed at other sinks and the overall effect of the voids is to act as vacancy sources. Grain boundaries and dislocations provide such sinks and it seems apparent that chemical as well as geometric and kinetic factors are important to account for the wide diversity of observations. Much scope thus remains to evaluate the operation and effectiveness of vacancy sinks.

3. VACANCY FLUXES IN CREEP

For materials in service at elevated temperatures, creep strength and creep fracture resistance may be improved unidirectionally at the expense of properties in the transverse directions (Arzt 1988). This is achieved by the development of an elongated grain structure oriented in the direction required to withstand the highest principal tensile stress. Such a grain shape can be maintained in mechanically alloyed materials in which grain boundaries are stabilised by pinning by fine, inert oxide particles. It is only recently that full evaluations of the three dimensional mechanical properties have been attempted. A full analysis of creep life is compounded by the separate but inter-related processes of creep deformation and fracture but in the present paper only the form of deformation will be considered.

A feature of the microstructural changes caused by creep in materials of this kind based on γ' hardened Nickel-base alloys is the occurrence of precipitate free regions adjacent to grain boundaries perpendicular to a tensile stress or parallel to a compressive stress (Arzt and Timmins 1990). Such observations are widely interpreted on the basis of a vacancy flux with grain boundaries acting as vacancy sources and sinks depending on their orientation with respect to the applied stresses (Burton 1977). The theory of this creep mechanism is well established and has recently been extended to predict creep rates of materials with non-equiaxed grains under multi axial stresses in the case where grain boundaries can act as perfect vacancy sources and sinks (Greenwood 1985). This implies that the time for vacancy emission and absorption is negligible compared with the time spent in diffusing through the grains. (Grain boundary diffusion can also be important in some circumstances but will not be considered here). Whilst this theory appears adequate to interpret the behaviour of pure metals and some simple alloys, evidence (Arzt and Timmins 1990) suggests that it is less successful in representing the properties of the complex mechanically alloyed materials. Some influence of grain shape however still remains and it seems likely that a modified theory must take into account some inhibition of the operation of vacancy sources and sinks.

It is not yet clear how such modifications may be incorporated and a rigorous analysis is not attempted here but the requirement to match several closely defined conditions imposes severe restrictions on the form of the equations by which anisotropic creep strength may be described. These conditions will be identified by first considering the case where vacancy sources and sinks are not inhibited.

3.1 Anisotropy of Creep Strength with Uninhibited Vacancy Sources and Sinks

This situation has been rigorously analysed for three unequal grain dimensions (Greenwood 1985) and the results can readily be applied to the simpler case of cylindrical symmetry corresponding to an elongated grain length L and a transverse dimension B. Now it is useful to define 'creep compliance coefficients' S_{ij} by analogy with the compliance coefficients widely used in describing anisotropic elasticity such that the creep rates $\dot\epsilon_i = \Sigma_{j=1}^{6} S_{ij}\sigma_j$ are substituted for elastic strain. Taking orthogonal axes and representing the three principal stresses by σ_x, σ_y and σ_z for σ_1, σ_2 and σ_3 with the shear stresses τ_{yz}, τ_{zx} and τ_{xy} for σ_4, σ_5, and σ_6 and denoting the shear strain rates $\dot\gamma_{yz}$, $\dot\gamma_{zx}$ and $\dot\gamma_{xy}$ respectively, a matrix formulation can be constructed. This would have 6 x 6 = 36 elements in a general case but the geometrical restrictions of cylindrical symmetry reduce to five the number of independent creep compliance coefficients that are required, so the matrix is written

$$
\begin{bmatrix}
\dot\epsilon_x \\
\dot\epsilon_y \\
\dot\epsilon_z \\
\dot\gamma_{yz} \\
\dot\gamma_{zx} \\
\dot\gamma_{xy}
\end{bmatrix}
=
\begin{bmatrix}
S_{11} & S_{12} & S_{13} & - & - & - \\
S_{12} & S_{11} & S_{13} & - & - & - \\
S_{13} & S_{13} & S_{33} & - & - & - \\
- & - & - & S_{44} & - & - \\
- & - & - & - & S_{44} & - \\
- & - & - & - & - & 2(S_{11}-S_{12})
\end{bmatrix}
\begin{bmatrix}
\sigma_x \\
\sigma_y \\
\sigma_z \\
\tau_{yz} \\
\tau_{zx} \\
\tau_{xy}
\end{bmatrix}
$$

Here the long grain dimension L is in the z-direction and the creep compliance coefficients take on the following relationships derived from the complete three dimensional analysis; $S_{11} = \alpha(L^2+B^2)/H$, $S_{12} = -\alpha L^2/H$, $S_{13} = -\alpha B^2/H$, $S_{33} = 2\alpha B^2/H$ and $S_{44} = 4\alpha/(L^2+B^2)$ where $\alpha = 12D\Omega/kT$, in which D is the self diffusion coefficient and Ω the atomic volume. $H = B^2(B^2+2L^2)$. It may further be noted that $2(S_{11}-S_{12}) = 2\alpha(2L^2+B^2)/H = 2\alpha/B^2$.

The above formulation gives a complete description of the diffusional creep process under any stress system in materials with an elongated grain structure in which grain boundaries can act freely as vacancy sources and sinks. There is however evidence that grain boundaries in complex mechanically alloyed materials are inhibited in their operation (Arzt and Timmins 1990) and some new proposals are made next to analyse this situation.

3.2 Possible Effects of the Inhibition of Vacancy Sources and Sinks

Much consideration has been given to the operation of vacancy sources and sinks at grain boundaries and the concept of vacancy emission and absorption from grain boundary dislocations has provided a useful basis (Ashby 1969). One conclusion from such analyses clearly indicates that when the emission and absorption steps are rate controlling, then creep rate is inversely proportional to grain size for equiaxed grains, in contrast with the inverse square relationship without source and sink inhibition. A more general analysis independent of any details of the emission and absorption process confirms this result (Greenwood 1970) when these processes form the rate controlling step and additionally points to the extension of the relationship in the form that the creep rate at constant stress and temperature in the direction of grain elongation is inversely proportional to grain length. Thus for $\sigma_z > 0$, $\sigma_x = \sigma_y = 0$ and L > B, then $\dot\epsilon_z \propto 1/L$. Another is that for $\sigma_x = -\sigma_y$, then $\dot\epsilon_z = 0$, independently of L when $\sigma_z = 0$. Two further requirements are that the creep rates are unaffected by hydrostatic pressure and secondly that creep occurs at constant volume in the absence of creep damage. These conditions are highly restrictive in the form of equation that allows them all to be simultaneously satisfied.

There is an additional complication in that interfacial control of vacancy emission and absorption leads to power law creep, (Arzt and Timmins 1990), with grain boundary dislocation theory suggesting that the stress exponent n = 2 (Ashby 1969). Independently of the magnitude of n however, conditions demand that anisotropic behaviour is governed by linear terms in stress such that creep under unidirectional stress σ_x takes the form $\dot{\epsilon}_x \propto \sigma_x \sigma_e^{n-1}$ where σ_e may be considered (Cottrell 1964) as a Von Mises equivalent stress given by $\sigma_e = [(\sigma_x - \sigma_y)^2 + (\sigma_y - \sigma_z)^2 + (\sigma_z - \sigma_x)^2]^{1/2}/\sqrt{2}$. All these conditions may be satisfied if $S_{11} = A\sigma_e^{n-1}(L+B)/B(B+2L)$, $S_{12} = -A\sigma_e^{n-1}L/B(B+2L)$, $S_{13} = -A\sigma_3^{n-1}B/B(B+2L)$, $S_{33} = 2A\sigma_e^{n-1}B/B(B+2L)$ and $S_{44} = 4A\sigma_e^{n-1}/(L+B)$ where A is a constant at constant temperature and the grain dimension L is along the z axis.

4. CONCLUSIONS

Two illustrations are given of the importance of a knowledge of the operation of vacancy sources and sinks. Both are complex in their details but present opportunities for a full modelling of vacancy behaviour in situations where significant macroscopic effects can occur. Some approaches are suggested in the interpretation of experimental results and in rationalising their description. Improvements in the two areas, one in the process of diffusion bonding and the other in the development of enhanced creep resistance through anisotropy could result from such considerations.

ACKNOWLEDGEMENTS

The author is grateful for useful discussions with his colleagues in the School of Materials of the University of Sheffield and for support from the Science and Engineering Research Council and the Leverhulme Trust Foundation.

REFERENCES

Arzt E 1988 *'New Materials by Mechanical Alloying Techniques,* Ed Arzt E and Shultz L, D G M Oberusel FRG p 185.
Arzt E and Timmins R 1990 *Int Conf on 'Structural Applications of Mechanical Alloying'* ASM, Myrtle Beach SC USA.
Ashby M F 1969 *Scr Metall* 3 837.
Balluffi R W 1954 *Acta Metall* 2 194.
Barnes R S, Redding G B and Cottrell A H 1958 *Phil Mag* 3 97.
Buckle H and Blin J 1952 *J Inst Met* 80 385.
Bueren H G van 1960 *'Imperfections in Crystals'* North Holland Publishing Co Amsterdam p 259.
Bullough R and Finnis M W 1983 *Conf on 'The Mechanics of Dislocations'* ASM Metals Park, Ohio, USA pp 11-19.
Burton B, 1977, *'Diffusional Creep in Polycrystalline Materials' Trans Tech Publications,* Aedermannsdorf, Switzerland.
Clay B D and Greenwood G W 1972 *Phil Mag* 25 1201.
Cottrell A H 1964 *'Mechanical Properties of Matter'* John Wiley and Sons Inc New York.
Greenwood G W 1970 *Scr Metall* 4 171.
Greenwood G W 1985 *Phil Mag* 51 537.
Harris J E, Jones R B, Greenwood G W and Ward M J 1969 *J Austral Inst Metals,* 14 154.
Kheder A R I 1976 *PhD Thesis University of Sheffield.*
Le Claire A D and Barnes R S 1951 *J Metals* 3 1060.
Rauh H and Bullough R, 1985 *Proc R Soc* (London) A397 121.
Rauh H, Hippsley C A and Bullough R 1989 *Acta Metall* 37 269.
Reed-Hill R E 1964 *'Physical Metallurgy Principles' Van Nostrand Reinhold,* New York.
Smigelskas A D and Kirkendall E O 1947 *Trans AIME* 171 130.

The Peierls-stress for various dislocation morphologies

R. Bullough, A.B. Movchan* and J.R. Willis+

Corporate Research Directorate, AEA Technology, Harwell Laboratory, Oxon. OX11 0RA

*School of Mathematical Sciences, University of Bath, Claverton Down, Bath. BA2 7AY; Permanent address: Institution of Precision Mechanical Engineering and Optics, 14 Sablinskaya St., Leningrad, USSR.

+School of Mathematical Sciences, University of Bath, Claverton Down, Bath. BA2 7AY

ABSTRACT: The Peierls-Nabarro model of a dislocation with its associated singular integral equation is famous for the analytic solution for the relative displacement and Peierls-stress it yields for a straight, isolated dislocation when a sinusoidal interatomic shear-law of force prevails across the glide plane of the dislocation. In this paper we demonstrate that such singular integral equations can be easily solved numerically for both arbitrary laws of force and for complex dislocation morphologies. The procedure is illustrated by obtaining, for the first time, an accurate solution of the Peierls-Nabarro model for a tilt boundary; the dependence of the relative displacement and Peierls-stress on the dislocation spacing in the boundary is obtained.

1. INTRODUCTION

The so-called Peierls-Nabarro model (Peierls 1940, Nabarro 1947) of a long straight edge or screw dislocation in an infinite crystalline medium is constructed by replacing the discrete crystalline medium by two semi-infinite linear elastic continua that are "joined" across the glide-plane of the dislocation by non-Hookean material which transmits a suitable (sinusoidal) interatomic shear-law of force. The presence of this non-Hookean material at the core of the dislocation thus eliminates the elastic singularity associated with the usual purely elastic 'cut and weld' dislocation and permits an, albeit somewhat primitive, estimate to be made of the atomic configuration in the dislocation core region. The model has the important conceptual advantage that the relative atomic displacement across the glide-plane, the associated misfit energy and the 'Peierls' force to cause the dislocation to move by glide can, at least for the sinusoidal interatomic law of force, be expressed in analytic form. The model thus provided the first rigorous demonstration that the shear stress required to move a dislocation was the order of a hundredth of the theoretical shear strength ($\sim$ shear modulus/20) of the host material.

For an isolated straight edge dislocation with a Burgers vector of strength b, the relative displacement across the glide-plane, $\Phi(x)$, is given by the integral equation

$$F(\Phi(x),b) = -\frac{1}{1-\nu} \int_{-\infty}^{\infty} \frac{\Phi(x')}{(x-x')^2} \, dx' \qquad (1)$$

where ν is Poisson's ratio and the x co-ordinate lies in the glide plane and the dislocation line is located at $x = 0$. This equation is singular and for physically sensible forms of the force-law F, is highly non-linear. Specifically when the force-law is sinusoidal:

$$F(\Phi(x),b) = F_s(T) = \sin(T), \quad T = 2\pi\Phi(x)/b \qquad (2)$$

equation (1) becomes the Peierls-Nabarro integral equation with the exact solution (Peierls 1940):

$$\Phi(x) = -\frac{b}{\pi} \tan^{-1} \left[\frac{2(1-\nu)x}{b}\right]. \qquad (3)$$

The Peierls stress, σ, to move the dislocation in a simple cubic host crystal with lattice parameter b is given by

$$\sigma = \max_{\alpha} \left[-\frac{1}{b}\frac{dE}{d\alpha}\right] \qquad (4)$$

where

$$E = \frac{b^2\mu}{8\pi^2} \sum_{n=-\infty}^{\infty} G\left[-2\pi\Phi\{(\alpha + \frac{n}{2})b\}\right] \qquad (5)$$

is the varying part of the misfit energy across the glide plane when the dislocation is displaced a distance αb and in general

$$G(x) = \int_{0}^{x} F(t) \, dt. \qquad (6)$$

For the sinusoidal law of force with $F = F_s$, given by (2), and thence $\Phi(x)$ by (3), the summation in (5) can be evaluated in closed form to yield the well known (approximate) analytic result for the Peierls-stress (4):

$$\sigma = \frac{2\mu}{1-\nu} \exp\left[-2\pi/(1-\nu)\right] \qquad (7)$$

where μ is the shear modulus of the (elastically isotropic) host material.

Any departure from the simple sinusoidal force-law (2) precludes equation (1) having an analytic solution, similarly, any change of dislocation morphology from the long straight isolated form leads to a change of kernel in the integral equation which again will usually preclude the existence of an analytic solution to the resulting integral equation even when the force-law is sinusoidal. The purpose of this present paper is to show, with the explicit example of a tilt boundary, that such singular equations can now be easily solved numerically by exploiting the representation of the Finite-Part integral (Willis and Nemat-Nasser 1990):

$$\int_{-A}^{A} \frac{Q(y)}{(y-x)^2} \, dy = \int_{-A}^{A} \frac{Q(y) - Q(x) - (y-x) \, Q'(x)}{(y-x)^2} \, dy$$

$$- \frac{2 \, A \, Q(x)}{A^2 - x^2} + Q'(x) \, \ln \left| \frac{A - x}{A + x} \right| \tag{8}$$

where we note that when Q is a smooth function the integral on the right-hand side of (8) is convergent. Full details of the numerical procedure with a more extensive discussion of applications will be published elsewhere (Bullough, Movchan and Willis 1991).

2. THE VERTICAL ARRAY OF DISLOCATIONS

Such an arrangement of edge dislocations constitutes a simple-tilt boundary. If h is the separation between the edge dislocations then the angle of tilt, θ, is given by $\theta = \tan^{-1}[b/2h]$. The relevant integral equation for $\Phi(x)$ that replaces (1) is (Bullough 1955, Bullough and Tewary 1979):

$$F(\Phi(x),b) = - \frac{\pi}{h \, (1-\nu)} \int_{-\infty}^{\infty} \Phi(x') \, \frac{d}{dx'} \, \frac{\pi(x-x')/h}{\sinh^2[\pi(x-x')/h]} \, dx'. \tag{9}$$

This equation has been solved numerically by exploiting the representation (8) for the sinusoidal law F_s and the 'more realistic' asymmetric law with the polynomial form

$$F(\Phi(x),b) = F_p(T) = \left\{ \begin{array}{l} R(T), \; T > 0 \\ -R(-T), \; T < 0, \end{array} \right\}$$
$$R(T) = T(\pi-T)(\pi - 0.7T)/\pi^2 \tag{10}$$

where $T = 2\pi\Phi(x)/b$. The force-laws (2) and (10) are shown in Figure 1 and the corresponding solutions for $\Phi(x)$ are shown in Figures 2 and 3 respectively for a range of h values. In the numerical results presented here the value of Poison's ratio has been taken equal to 0.3 and all lengths are expressed in units of b. When the dislocation spacing in the array $h \to \infty$, the integral equation (9) reverts to the Peierls-Nabarro form (1) for the isolated edge dislocation; the variation of $\Phi(x)$ with the sinusoidal force-law, for $h = 100b$ ($\theta \sim 0.28°$) in Figure 2 is

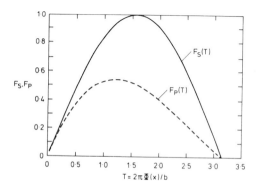

Fig. 1. The interatomic shear-laws of force $F_s(T)$ and $F_p(T)$, given by (2) and (10) respectively, used in the calculations.

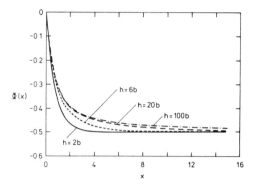

Fig. 2.   The change in relative displacement Φ across the glide-plane of an edge dislocation in a tilt-boundary, with dislocation separation h: for the sinusoidal force-law F_s (equation (2)).

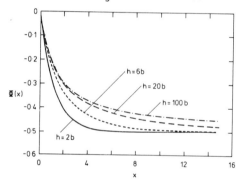

Fig. 3.   As in Figure 2: for the polynomial force-law F_p (equation (10)).

indistinguishable from the analytic solution (3). This confirmation of the accuracy of the numerical solution has been further reinforced by solving the corresponding integral equation for the vertical array of screw dislocations

$$F(\Phi(x),b) = -\frac{2\pi}{h} \int_{-\infty}^{\infty} \Phi(x') \frac{d}{dx'} \left[\coth \left[\pi(x-x')/h \right] \right] dx' \qquad (11)$$

when F has the sinusoidal form (2); this equation has the exact analytic solution (Bullough 1955)

$$\Phi(x) = -\frac{b}{\pi} \tan^{-1} \left[\frac{\tanh\ (\pi x/\varepsilon)}{\tan\ (\pi\gamma/\varepsilon)} \right] \qquad (12)$$

where ε and γ are given by the simultaneous equations:

$$h = \varepsilon - 2\gamma, \quad \sin\ (2\pi\gamma/\varepsilon) = b\pi/\varepsilon. \qquad (13)$$

Since this vertical array of screw dislocations is somewhat unphysical, being unstable and creating long range shearing of the host lattice, there is little point in presenting the actual computed form of $\Phi(x)$. Suffice it to say that the computed solutions of (11), for a range of h, were indistinguishable from the analytic solution (12) when F was chosen to be sinusoidal (equation (2)).

The solutions of (9) for the vertical array of edge dislocations for the two force-laws have been used to evaluate, from equation (4), the Peierls-stress to move the array. The results are given in Figures 4 and 5 for the force-laws (2) and (10) respectively; the absolute accuracy of these computed small stresses is not high but systematic errors are negligible and therefore their variations with h are reliable.

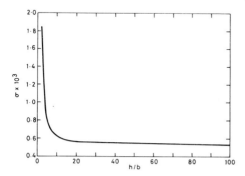

Fig. 4. The computed dependence of the Peierls stress for a tilt-boundary on the dislocation separation h: for the sinusoidal force-law F_s (equation (2)).

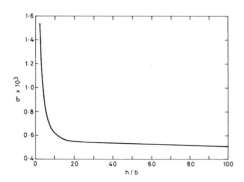

Fig. 5. As in Figure 4: for the polynomial force-law F_p (equation (10)).

3. DISCUSSION

The variations of $\Phi(x)$ with dislocation spacing for the tilt boundary, shown in Figures 2 and 3, are of considerable interest. They indicate that the Peierls "widths" of the individual dislocations do contract as the tilt angle increases and the strong effect this contraction has on the mobility of such boundaries is manifest in the significant concomitant variation of the Peierls-stress shown in the corresponding Figures 4 and 5. Needless to say considerations of the mobility of interfaces have also relevance to more general interfaces such as those associated with martensitic transformations and the fact that atomic core variations with orientation should occur, as demonstrated by the present calculation, which can in turn strongly affect the interface mobility, must be taken account of in any glissile surface dislocation representation of such transformations (Bullough and Bilby 1956).

As was emphasised in Section 1, singular integral equations of the Peierls type can now be easily solved numerically. We are, in fact, in the process of investigating various dislocation morphologies including the glissile dislocation loop and dislocation dipoles. Our results for the loop, which has an associated integral equation with a complex kernel involving combinations of elliptic functions plus an identical singularity to (1), demonstrates that their glissile mobility increases as their radius decreases, in attractive contrast to the mobility variation of the tilt boundary!

Finally, the evaluation of an accurate solution of the Peierls-Nabarro tilt boundary, albeit for a simple cubic lattice, gives particular pleasure to one of us (RB) since he derived the integral equation (9) as a research student, could not solve it and the present accurate solution has taken $\sim$ 36 years to appear; better late than never!

REFERENCES

Bullough R 1955 Ph.D. Thesis (Sheffield University, Sheffield).
Bullough R and Bilby B A 1956 Proc. Phys. Soc. Lond. B69 1276.
Bullough R, Movchan A B and Willis J R 1991 To be published.
Bullough R and Tewary V K 1979 Dislocations in Solids Vol. 2 -
   Dislocations in Crystals (FRN Nabarro Ed: North-Holland) pp. 1-65.
Nabarro F R N 1947 Proc. Phys. Soc. Lond. 59 256.
Peierls R 1940 Proc. Phys. Soc. Lond. 52 34.
Willis J R and Nemat-Nasser S 1990 Quart. J. of Appl. Maths. XLYIII(4)
   741.

The texture transition and shear bands in f.c.c. metals

C S Lee*, B J Duggan* and R E Smallman+

*Department of Mechanical Engineering, University of Hong Kong;
+School of Metallurgy and Materials, University of Birmingham, Edgbaston, Birmingham, B15 2TT, U.K.

ABSTRACT   The FCC rolling texture transition is briefly reviewed and problems in existing theory are highlighted. Deformation banding is observed to be important and its inclusion into modelling in a Taylor framework produces predictions closer to experiment. {110}<112> is formed progressively in the new model and {123}<634> should be considered as an important rolling texture component in both high and low SFE materials. Consideration is given to shear bands forming in a catastrophic yield-like manner.

1. THE TEXTURE TRANSITION

By 1952 Beck and his coworkers had established that the textures formed in aluminium, copper and α brass were radically different, the brass texture being centred on {110}<112> and the "pure" metal texture a complex orientation distribution now described as {112}<111> + {321}<634>. The description of FCC rolling textures as "alloy" or "pure" metal type seemed natural until Smallman (1955-56) in extensive early work at Harwell showed that the "alloy" type texture was formed in silver when rolled at room temperature. He also showed that alloying any of the common FCC metals, other than silver, with elements which entered into solid solution produced a transition from the pure metal to the alloy texture dependent on solute misfit, valency etc. The Harwell work was also instrumental in establishing that the rolling temperature was a vital element in determining the kind of texture formed, low temperatures favouring the brass type and higher temperatures the copper type. The result obtained for silver was difficult to accept for it was believed at that time that silver and copper had similar stacking fault energies (SFE) and so other explanations were given. Smallman and Westmacott (1957) measured the S.F.E.'s of F.C.C. metals and alloys to determine the relative values but it was not until a higher purity silver sample was rolled by Hu and Cline (1961) that the {110}<112> texture was accepted as being characteristic of the metal itself. Finally Smallman and Green (1964) demonstrated conclusively that the texture formed in an FCC metal or alloy at a temperature no higher than 0.25 Tm was a function of the SFE, high SFE metals and alloys giving the {112}<111> + {123}<634> texture and low SFE metals and alloys {110}<112>.

Theories of the rolling texture transition fall into two groups, the first postulates that normal octahedral slip leads to the development of the brass texture {110}<112> and that additional deformation processes in copper such as non-crystallographic shear, cube slip, cross-slip and dislocation interactions modify the copper texture. The second group assume that octahedral slip leads to the copper texture while deviations to the brass texture involves other mechanisms e.g. latent hardening and overshooting or twinning. Wassermann (1963) made the critical contribution that normal slip in both copper and brass produces similar textures centred on {112}<111> and {110}<112> at 40% - 50% reductions, but mechanical twinning in {112}<111> crystals, reorients them to {552}<115> which then rotates by octahedral slip to {110}<112> via {110}<001>. This picture remained substantially unchanged until interest focussed in the 1970s on shear bands which form in brass after ~50% rolling and increase in number and volume as strain increases. Detailed work by Duggan et al. (1978) on shear banding has lead to a modified version of the Wassermann theory. The five stages identified in the development of the brass texture now widely accepted are: (i) 0-50% reduction - deformation occurs by {111} slip in both copper and brass, producing similar copper-like textures; (ii) 40 - 60% reduction - extensive twinning occurs in brass in orientations close to {112}<111>; twins are fine, 0.02 - 0.2 μm thick and represent ~25% of the volume of twinned regions; (iii) 50 - 80% reduction - in twinned volumes slip is restricted by the twin boundaries to planes parallel to those boundaries leading to overshooting such that orientations of the type {111}<uvw> are formed by coupled rotation; the component is detected in pole figures; (iv) 60 - 95% reduction - shear bands divide heavily twinned volumes, destroying {111} components. Coupled rotation and shear banding compete to produce a maximum 111 intensity in the range 80 - 90%, after which it declines. Shear bands are composed of crystallites of 0.02 - 0.3 μm in diameter in sheets ~1-2 μm wide, and there is some evidence of {110}<001> in the bands; (v) above 85% reduction - homogeneous slip processes in the increasing volume of shear band material is believed to lead to the stable {110}<112> texture.

2.   PHYSICAL MODELLING OF THE TEXTURE TRANSITION

Computer modelling of texture development assumes plane strain deformation and application of the Taylor principle, which assumes a unique strain tensor equal to the macroscopically imposed strain. This ensures strain compatibility throughout the polycrystal, but neglects stress equilibrium, and requires the selection of slip systems which, in combination, gives minimum internal work to accommodate the imposed strain. For FCC metals there are twelve slip systems and for any arbitrary shape change five systems are required. Ambiguities arising in the choice of slip systems have been considered by various workers and texture simulations based on these approaches all give similar results, but not identical to the measured copper texture (Hirsch and Lücke 1989b). A spread of orientation from {4,4,11}<11,11,8> to {123}<634> is predicted but not {112}<111> and {110}<112>. Incorporation of twinning in the Taylor modelling produces a brass texture, but different physical criteria for allowing twinning to occur produces identical results, none of them consistent with the microstructural evidence.

An improvement in modelling accuracy recognised that the grains elongate and become flatter as rolling proceeds, which has the effect of reducing certain misfit strains between neighbouring crystals to a point where they can be neglected. This relaxed constraint (RC) Taylor model has been applied to texture prediction by various workers with the effect that {112}<111> and {123}<634> are both present in the simulated copper texture. A further development is the incorporation of both the FC Taylor theory at low strains and the RC Taylor theory as grains become flat at high strains, with some improvement in the predictions, but {110}<112> is still not predicted. This component, which comprised some 20% of the copper texture, can be formed in computer models only if another constraint, not allowed by grain geometry, is permitted. Of course, allowing relaxation in this manner, destroys the stability of other components including the major components {112}<111> and {123}<634> (Hirsch and Lücke 1989b). These problems, plus the fact that the textures formed in computer simulations are too sharp and form more rapidly with strain than they do in reality, are problems which are being actively researched. However, modelling procedures based on normal slip within the Taylor framework have reached their limit without the introduction of a different deformation mode. In effect this new mode has been adduced without justification, to explain the formation of {110}<112>, i.e. allowing a strain relaxation which does not seem to be intuitively reasonable, given the constraints. Recent work (Lee, 1991) shows that a likely deformation mode is deformation banding, whereby certain crystals in a polycrystalline aggregate subdivide and rotate in different directions. The process is orientation dependent in cubic metals and the boundaries can be either sharp or diffuse. Different regions can carry different strains provided the work done by slip within the bands is less than that done for homogeneous deformation and provided the bands can be arranged such that the nett strain matches the overall deformation.

3. THE COPPER ROLLING TEXTURE

3.1 Experimental Observations

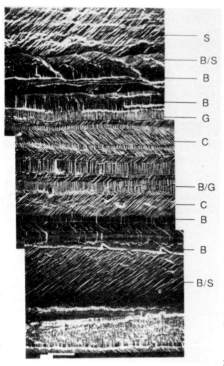

Figure 1 shows a SEM micrograph from the longitudinal section of a coarse grained copper specimen after etching by the Köhlhoff technique in which {111} planes of copper alloys are most slowly attacked and hence revealed. Layers 1 to 50 μm thickness are observed, the orientations of which are typical of rolled copper, i.e. {123}<634> or S, {110}<112> or B, {110}<001> or G and {112}<111> or C. Boundaries between the layers are often crystallographically sharp. What is remarkable about this micrograph, however, is the lack of correspondence between the original grain size and the layer thickness after rolling 85%. The initial grain size was 3000 μm and the average grain thickness after 85% reduction should be ~450 μm but the average layer thickness was measured at four locations on two different specimens and found to be ~17 μm. This means that each grain has broken up to give ~25 layers of distinct orientation. Examination of the rolling plane section revealed the three dimensional nature of deformation banding and supports the idea that each grain has subdivided by deformation banding into a large number of elements of different orientation, the average width of each band being 120 μm. Hence the 3D structure is that of laths 120 μm wide and 17 μm thickness. Orientation relationships allow twinning to be eliminated as a formation mechanism. Copper with a grain sizeof 50 μm was also prepared, rolled to the same strain and etched. On average each grain was split 2.5 times in the LS, so there is some evidence of a grain size dependence on deformation banding. Textures showed a higher concentration of {112}<111> in the coarse grained material.

Fig. 1 Microstructure of cold rolled copper.

3.2 Modelling the Copper Texture to include Deformation Banding

Within the Taylor framework deformation banding in a particular grain requires the sum of the strains in different bands to equal the macroscopic strain. From a purely geometrical standpoint there can be six modes of deformation banding, shown in Figure 2. In each of the six modes, different shear deformation components can be non-zero in individual deformation bands provided the external conditions are met. Operation of all six deformation banding modes would relax all six shear components but be too complicated. Deformation banding of modes II, IV and VI are more favoured on energy grounds (Lee 1991) and shear strains ε_{31} and ε_{21} are easily accommodated by grain shape while ε_{32} is not. Hence a relaxed constrains approach could be used for ε_{31} and ε_{21}, and deformation banding for ε_{32}. This means allowing banding according to mode II, figure 2, to occur first, i.e., before ε_{31} and ε_{21}. The other components of shear strain which could also be relaxed by deformation banding, are instead treated as in the classical relaxed constraints model.

The constraints on deformation determines the number of slip systems which must be used in the deformation banding process, and it is possible to demonstrate formally (Lee, 1991) that provided volume is conserved and strain components ε_{21} and ε_{31} are relaxed because of grain shape, then only two independent constraints exist. This means, in principle at least, that only two slip systems are needed to accommodate the shape change. It must be stressed, however, that the relaxation of shear by deformation banding is not as free as in the RC model. This is because it is a condition for deformation banding to occur that the average shear strain over the grain should be zero, which implies that shear components in different regions of the grain must cancel each other. The two operative slip systems are selected according to two rules. The first is that the constraints conditions discussed above are fulfilled, and the second that the internal work done should be minimum. Once this has been achieved the crystal rotations can be calculated.

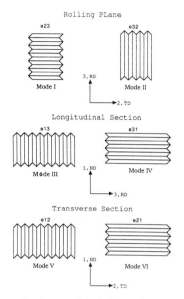

Figure 2 : Six possible deformation banding modes.

A random array of 100 grains was used as the computer material and textures computed after 30, 54, 79 and 92% equivalent rolling reduction, the strain in each step being 0.05. It is unrealistically assumed that all grains deform by deformation banding. Figure 3 shows the measured texture of copper rolled 75% obtained by Hirsch and Lücke (1989a) using the Euler space projection, superimposed on the computed textures based on deformation banding. The predictions are fairly accurate and superior to the older models. (1) All the components are predicted which is not true of the FC and RC Taylor models. (2) The predicted components are closer in orientation to the experimental peaks, in particular the component {4,4,11}<11 11 8> is sparsely populated. In the other models special techniques have to be used to rid the simulation of this component . (3) The predicted texture has wider spread than other model textures and in this respect is closer to experiment. (4) All other modes use more slip systems than are observed to operate. The deformation banding model requires fewer, and so is closer to experiment.

Nevertheless, the model unrealistically assumes that all grains deform by deformation banding and the orientation-dependence of deformation banding is not considered. For example the {123}<634> orientation is forced to divide by deformation banding in the computer model but the orientations are rather similar and remain so until 87% reduction. Only after 92% does the orientation spread. The behaviour of {100}<001> oriented grains is typical of another set of orientations, in which deformation banding produces divergent rotation paths upon further deformation. The modelling therefore overestimates the effect of deformation banding in orientations such as {123}<634>. In reality, grains of this orientation would deform by conventional relaxed constraints processes rather than by deformation banding and thus would be stable. Deformation banding in the computer makes {123}<634> unstable at high strains and hence underestimates its contribution to the final texture.

4.   ROLLING TEXTURES OF LOW SFE MATERIALS

The accepted model for the development of texture in α-brass outlined in section I is deficient in a number of ways, the most important concerning the behaviour of the orientations {110}<112> and {123}<634> at intermediate (i.e. 50% - 80%) rolling reductions. TEM studies reveal the presence of both components in significant amounts over this strain increment, and twinning is rare in (110)<112>. Thus a satisfactory deformation banding model, for {110}<112> does not involve twinning at intermediate strains, which represents a significant departure from the Wassermann theory that requires twinning to account for its origin. For the sake of clarity, individual components of the texture are considered in isolation, together with the specific deformation mode thought to be important in its formation.

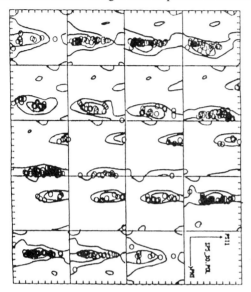

Figure 3 : Simulated fcc textures after 79% rolling with the deformation banding model.

The {110}<112> orientation develops progressively when deformation banding is allowed, as already discussed for copper. This means that in principle, twinning and shear banding are not required for its formation. Hence its presence at low strains in brass can be explained by deformation banding. However, homogeneous slip in the fine crystallites was suggested as the mechanism considered to be responsible for the tremendous strengthening at high strain. This idea is developed further by noting that Chung et al. (1988) found extensive shearing to be occurring in this material, even though etched shear bands did not appear. It is suggested that homogenous shear deformation of the fine crystallites in addition to homogeneous plane strain, produces {110}<112>.

Crystals of {112}<111> orientation twin according to Wassermann and by a coupled rotation of both matrix and twin elements align with of <111> parallel to ND which is subsequently destroyed by shear banding. There is some evidence that {110}<112> and {110}<001> are formed in the shear bands (Duggan et al., 1978).

TEM studies show that there is a considerable volume of {123}<634> in brass after 85% (Lee 1991) even though this component is not considered to be part of the brass texture. Indeed in a simulated 111 pole figure comprising four ideal orientations made up of 50% {110}<112>, 15%{123}<634>, 25% of material oriented between these positions and 10% {110}<001>, the pole figure is recognisably brass-like but the {123}<634> and the orientations near to it are not obvious. If it is accepted that these are formed in the brass texture, the behaviour of these components has to be considered. In particular, {123}<634> twins to an orientation close to itself, and so is not destroyed by this process. Deformation banding likewise produces an orientation close to {123}<634>, and only after high strains (~90%) does it rotate away to {110}<112>. There is some evidence that {123}<634> twins less frequently and at higher strains than {112}<111> and these twinned grains behave similarly to {112}<111>. It is thus clear that {123}<634> is rather stable in brass.

5. SHEAR BANDS

Shear bands are probably less important in the texture transition than previously thought. However the occurrence and characteristic microstructure has been extensively studied. Dillamore et al. (1979) have related shear band formation to the general instability condition

$$\frac{1}{\sigma}\frac{d\sigma}{d\varepsilon} = \frac{n}{\varepsilon} + \frac{m}{\dot{\varepsilon}}\frac{d\dot{\varepsilon}}{d\varepsilon} + \frac{1+n+m}{M}\frac{dM}{d\varepsilon} - \frac{m}{N}\frac{dN}{d\varepsilon} \leq 0$$

where M is the Taylor factor, m the strain-rate sensitivity, n the work hardening index and N the mobile dislocation density. The 1/M dM/dε term corresponds to geometrical softening such that instability is favoured if it causes a lattice rotation into a geometrically softer orientation. The term in dN/dε has been neglected in this context but may be appropriate when shear bands form in a catastrophic yield like manner (Lee and Duggan, 1990).

For a finely twinned grain width w with a potential shear band thickness t at an angle ß to the twin lamellae, n_p dislocations may pile-up against the twin boundaries and at a critical stress break through and propagate as a shear band using polyslip. The density of dislocations freed is given by $dN = (n_p/tw)\sin\beta$ and the strain $d\varepsilon = \Sigma|\gamma_i|/M$ where the summation is done over the active slip systems in the shear band. The shear strain on the ith system is $\gamma_i = n_{pi}b/(t\sin\xi_i/\sin\beta)$ where ξ_i is the angle the slip system makes with RD, so that $n_{pi} = \gamma_i t|\hat{R}.\hat{n}_i|/b\sin\beta$. With $n_p=\Sigma n_{pi}$ then $dN = (1/bw)\Sigma|\hat{R}.\hat{n}_i\gamma_i|$ and the yield term in the instability equation becomes $(m/Nwb)\Lambda$.

The dislocation avalanche factor Λ is related to shear band initiation but not propagation. It is naturally negative but the larger its value the more likely is instability; appropriate values for the parameters give $m\Lambda/Nwb$ ~unity when the instability condition is zero and hence N has the reasonable density ~3 x 10^{13} m^{-2}. The shear band angle is determined by some minimum energy principle but the value of Λ along with $1/M$ $dM/d\varepsilon$ must be such as to meet the instability criterion and trigger the band. In this sense Λ and the propagation angles are related. Calculation shows that Λ is near to maximum at 35° in $(111)[\bar{1}\bar{1}2]$ where the effective Taylor factor M' is minimum, and only two slip systems are needed for shear at ~35°. These conditions are not met for a positive shear band in the twin related orientation $(111)[11\bar{2}]$.

6. CONCLUSIONS

The FCC rolling texture transition cannot be explained by a simple model. The influence of temperature and stacking fault energy on the mechanisms of plastic deformation leads to different deformation modes and divergent deformation pathways.

The copper texture has been produced in a Taylor framework by the inclusion of deformation banding into the modelling procedure with simplifying assumptions. All the important components are predicted and develop with strain similar to experimental.

The rolling texture of brass is also better explained by the incorporation of deformation banding to account for {110}<112> volumes. However, the rapid increase in {110}<112> with strains above 90% is due to shear within shear bands. The {1$\bar{1}$2}<111> component does twin but then behaves according to the later theory of coupled rotation. TEM shows that {123}<634> has considerable stability and should be considered as one of the brass texture components.

ACKNOWLEDGEMENTS

The authors thank the Croucher Foundation for the provision of electron microscope facilities and studentship support (C.S. Lee).

REFERENCES

Chung C Y, Duggan B J, Bingley M S and Hutchinson W B 1988 Proc. 8th Int. Conf. on Strength of Metals and Alloys **1** 319
Chin, G.Y. 1969, Textures in Research and Practice (Springer-Verlag, Berlin) 236.
Dillamore, I.L., Roberts, J.G. and Bush, A.C., 1979, Metal Sci. <u>13</u>, 73
Duggan B J, Hatherly M, Hutchinson W B and Wakefield P T 1978 Met. Sci. **12** 343
Hirsch J and Lücke K 1989a, b, Acta Metall. **36** 2863, 2883
Lee C S 1991 Ph.D. Thesis, University of Hong Kong
Lee, C.S. and Duggan, B.J. 1990, ICOTOM <u>9</u>
Smallman R E 1955-56 J. Inst. Metals **84** 10
Smallman R E and Green D 1964 Acta Metall. **12** 145
Smallman R E and Westmacott K H 1957 Phil Mag. **2** 669
Wassermann G 1963 Z. Metallk. **54** 61

Simulation of steady state dislocation configurations

A. N. Gulluoglu* and C. S. Hartley**

*Postdoctoral Fellow, Department of Mechanical Engineering,
Florida Atlantic University

**Dean, and Professor of Mechanical Engineering,
College of Engineering, Florida Atlantic University
Boca Raton, FL 33431-0991 U.S.A.

ABSTRACT: A computer simulation technique to simulate numerically the formation and evolution of dislocation structures formed by the action of multiple slip systems as a function of interaction forces and applied stresses has been previously developed by the authors. In the present study, in addition to dislocation interactions and applied stresses, the interaction of dislocations with voids is considered and the effect of voids on the evolution of dislocation structures is investigated as functions of void densities and magnitudes of applied stresses. The resulting spatial distribution of dislocations is analyzed.

1. INTRODUCTION

Dislocation structures produced by deformation have been the subject of considerable interest for understanding the utility and mechanical properties of metals. The formation of dislocation cells and subgrain structures is of great importance to the understanding of deformation in single crystals as well as in polycrystalline solids. Dislocation structures depend sensitively on conditions present during deformation and on the presence of crystal defects or inhomogeneities in solids. Since crystal defects have stress fields associated with them, they interact with dislocations and cause changes in dislocation structures. The nature of dislocation microstructure formation and evolution in the stress-applied state remains cloudy. The difficulty is that essentially all experimental observations are made post-deformation. A complete understanding of mechanical behavior of metals is possible only by studying the actual dislocation structures produced during deformation taking into account the interaction between dislocations, applied stresses and crystal defects.

Recently, Gulluoglu (1991) introduced a computer simulation model based on molecular dynamics for studying dislocation structures in multiple glide. Formation and evolution of dislocation patterns formed by the action of multiple slip systems had been studied by Gulluoglu as a function of interaction forces, monotonic and cyclicly applied stresses, dislocation densities and temperatures. In this study, we not only consider the interactions of dislocations with all other dislocations present and with externally applied stresses, but also the interactions of dislocations with holes. These holes can be considered as vacancy clusters produced by irradiation of metals. This work examines the effects of hole density and strength on dislocation structures. The strength of the hole is given by $\sigma_{rr\,\infty}a^2$, where $\sigma_{rr\,\infty}$ is the applied stress at $r = \infty$ and a is radius of the hole.

In the following section, a simulation model developed for studying dislocation microstructures based on molecular dynamics is summarized. Results of numerical simulations are presented and analyzed in Section 3.   Finally, Section 4 presents a discussion of the simulation results with an analysis in terms of the dislocation pair distribution function and a comparison with experimental observations.

2. THEORETICAL BASIS OF SIMULATIONS

The mathematical formulation for computer simulations of dislocation structures is a many body problem.   In the present model, the "bodies" are dislocations, assumed to be straight, infinitely long and parallel, and to have mixed character. Mixed dislocations can move both in the direction of the edge component of their Burgers vectors and normal to their slip plane by glide and climb, respectively. Under these conditions, the simulation is essentially two-dimensional. In this model, we restrict our attention to cases where temperature is so low that climb can be neglected and the motion of dislocations is restricted to glide. In the present study, we employ orthogonal and coplanar slip systems. In this model, we have four slip systems with Burgers vectors parallel to [111] , [1$\bar{1}$1]   and [11$\bar{1}$], [1$\bar{1}\bar{1}$] on two orthogonal planes,(10$\bar{1}$)and (101), simulating slip in a body-centered cubic crystal. Dislocation lines lie parallel to the [0$\bar{1}$0] direction. Dislocations are represented by the symbols ⌐,⌐, ⌐,⌐ and ⌐, ⌐, ⌐,⌐ which indicate the signs and directions of Burgers vectors on the two orthogonal slip planes.

The simulation accounts for externally applied stresses, the interaction between dislocations, and the interactions between holes and dislocations. The total interaction force on one dislocation due to all the others is found in terms of its Burgers vector, **b**, and sum of the stress tensors of all other dislocations. The Peach-Koehler (1950) equation gives the force per unit length exerted on dislocation i by dislocation j,

$$f^{i\text{-}j} = \sum_{i \neq j} \left(\mathbf{b}_i . \sigma_j \right) \times \xi_i$$

(1)

where σ_j is the stress field of dislocation j evaluated at the position of dislocation i, ξ_i is a unit vector parallel to the line direction of dislocation i, and $\mathbf{b}_i$ is the Burgers vector of dislocation i.

Since crystal defects or inhomogeneities have stress fields associated with them, they interact with dislocations. Holes are typical of such defects. The stress field set up by cylindrical holes under biaxial stress is given by McClintock and Argon (1965). In the model, cylindrical holes are assumed to lie parallel to the dislocation lines. The stress field of the hole creates forces in the glide and climb directions of dislocations, which are determined by the Peach-Koehler formula, Eq. (1), in a manner analogous to the calculations described above for the dislocation interaction force case.

The stress field of dislocations and cylindrical holes decays very slowly, therefore dislocation-dislocation and hole-dislocation interaction cut-off distances are extremely important. Improper choices for interaction cut-off distance can introduce nonphysical effects in the simulation. Thus long range interaction forces must be included. The problem of truncation of the interaction force is eliminated by using periodic boundary conditions. Details of the problem of truncation and the solution to this problem using periodic boundary conditions is discussed in detail by Gulluoglu (1991). The stress field of an infinite vertical array of holes is obtained by summing up the stresses of individual holes using the method developed by Cottrell (1953) in a manner similar to that described by Gulluoglu (1991) for the case of an infinite dislocation array. Since the stress field of such a vertical array of holes decays exponentially, only a few arrays were included in the horizontal direction to account for stress fields of holes. This method for calculating the stresses includes all significant interactions and no interaction cut-off distance is required. By substituting the stress field of infinitely long dislocation (hole) arrays for the stress field of a single dislocation (hole) in Eq. (1) and letting $\mathbf{n}_i$ be the normal to the slip plane of dislocation i, the glide force due to all the other dislocations and holes f^g, can be obtained from $f^g = f. (\mathbf{n}_i \times \xi_i)$, where f is the total force per unit length due to dislocations and holes.

An applied external stress on a crystal may result in the glide and/or climb of dislocations within. External stresses create forces on dislocations, which can be determined by applying the Peach-Koehler formula, as described above (Eq. 1). In addition to these interactions, a damping force needs to be considered. It is well known that dislocation motion is a very energy dissipative process. A damping force (Hirth and Lothe 1968), similar to dynamic friction, is proportional to the velocity of the dislocation by a factor, Γ^g, called the glide damping coefficient.

Total glide forces on each dislocation are obtained by summation of the dislocation interaction forces, external forces, interaction with holes, and damping forces to give

$$f_i^g = \sum_{j \neq i} f_{ij\ int}^g + f_{i\ ext}^g + \sum f_{hole}^g - \Gamma^g \dot{x}_i \qquad (2)$$

where f_i^g is the glide force per unit length and x is the velocity in the x direction, and ($\cdot$) indicates differentiation with respect to time. The motion of dislocations subjected to these forces are solved in the framework of Newtonian mechanics with the Euler mid-point method (Gould and Tobochnik 1988). In this investigations, annihilation of dislocations of opposite sign is not considered, since it depends on the relative ease of dislocation climb. The equation of motion is non-dimensionalized in order to avoid possible numerical problems and to obtain a more general solution. Magnitudes of the constants have been chosen to represent α-iron(Weast 1975): μ=8.6x10^{10} N/m^2, b=2.482x10^{-10} m, c_t=3240 m/s, ν=0.291. The shape of computational cell was a 1 μm square.

It is characteristic of dislocation arrays that each dislocation tends to establish an ordered arrangement of its neighbors. Since the present simulations are two-dimensional and the stress field of a mixed dislocation is anisotropic, the environment of any dislocation can be characterized by a two-dimensional distribution function (Gulluoglu 1991). The dislocation distribution function displays most of the pronounced features of dislocation structures, such as periodicity in and normal to the slip plane. Therefore, two-dimensional distribution functions are the most appropriate method for analyzing the dislocation structures.

3. NUMERICAL SIMULATION AND RESULTS

Interaction forces on dislocations due to the stress fields of holes under biaxial stress results in alteration of dislocation structures. Stress fields around holes are created when stress is applied to a structure containing holes. The magnitude and character of the applied stress determines the strength of the hole. The effect of hole density and strength on dislocation motion has been investigated by means of the present simulations. In this investigation, hole diameter was fixed at 2x10^{-10} m, which is a typical void diameter observed in irradiated metals. In the first set of simulations, the effect of hole strength has been studied by using 3 different magnitudes of biaxial tension with a hole density of 10^{14} m^{-2} and a dislocation density of 3.2x10^{14} m^{-2}. In these simulations, the dislocation density is kept constant so as to isolate the effect of holes. The effect of hole strength on the geometrical arrangement of dislocations during deformation is shown in Fig. 1.(a-c). In this and the following figures, holes are represented by the symbol ●, dislocations are represented by the symbols ⊿, ⌐, ↖, ⊤, ◢, ⌐ and ⅂, ⌐, corresponding to positive and negative Burgers vectors. Simulation results of dislocation microstructures and the corresponding dislocation pair distribution functions are displayed in Fig. 1.(a-c). The distribution functions displayed in these figures have been averaged over dislocation structures from four set of simulations.

The effect of alterations in the density of holes at constant hole strength has been investigated for three different hole densities of 10^{12}, 10^{13} and 10^{14} m^{-2} under biaxial tension of σ/μ =5x10^{-3}. The effect of hole density on the dislocation structure and the corresponding dislocation distribution function are shown in Fig. 2.(a-c).

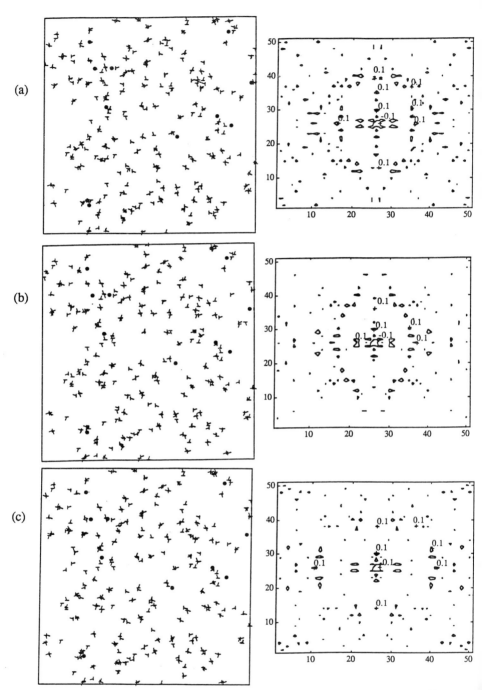

Fig. 1. Instantaneous dislocation structures after long simulations and corresponding contour plots of the distribution functions. Figure a-c correspond to deformation with biaxial tension of $\sigma/\mu = 10^{-3}$, 5×10^{-3} and 7.5×10^{-3} and with a hole density of 10^{13} m^{-2} and a dislocation density of 3.2×10^{14} m^{-2}.

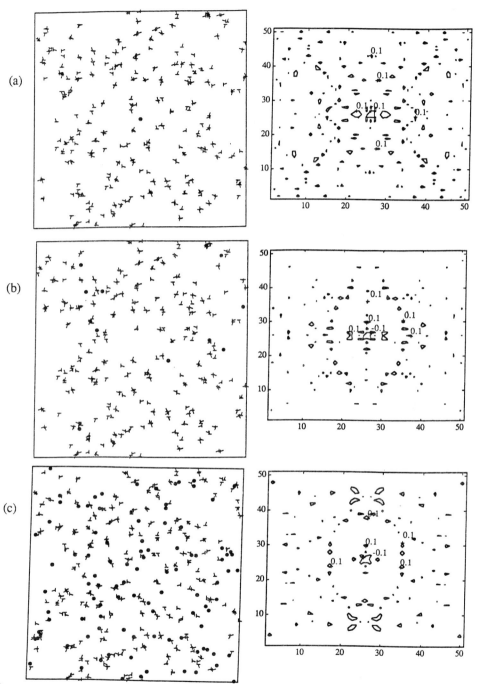

Fig. 2. Instantaneous dislocation structures after long simulations and corresponding contour plots of the dislocation distribution functions. Figure a-c correspond to deformation with hole densities of 10^{12}, 10^{13} and 10^{14} m^{-2} under biaxial tension of $\sigma/\mu = 5 \times 10^{-3}$ and a dislocation density of 3.2×10^{14} m^{-2}.

4. DISCUSSIONS AND CONCLUSIONS

The presence of dislocations in a crystal makes plastic deformation possible at stress levels far below the theoretical strength of the material. On the other hand, works of Maddin and Cottrell (1955) show that small amount of vacancies in the form of clusters or small holes in the matrix causes hardening by exerting a force on the dislocations. Fig. 1.(a-c) show the effect of hole strength on the dislocation motion during deformation. Close examination of dislocation structures and the corresponding distribution functions shows that reactions between dislocations on orthogonal slip planes result in barriers; coplanar dislocations are trapped at barriers, which produces dislocation pile-ups. Pile-ups of dislocations of like sign tend to line up in the direction normal to the slip plane, which forms walls, and dislocations having unlike sign tend to form dipoles along directions making an acute angle of 45° with the slip plane. The dislocation distribution functions show the presence of dislocation walls of finite extent and periodicity along the direction perpendicular to the these walls, which indicate the presence of a cell-like structure.

At low biaxial stresses, hole strength is lower, however the stress field helps the dislocations overcome small energy barriers in the dislocation pile-ups. Therefore the structures found at low stresses show more cell-like structures. Clearly, dislocation interaction forces dominate the dislocation motion. At high biaxial stresses, hole strength becomes more effective on dislocations. Increases in hole strength increase the resistance to dislocation motion. High hole strength prevents the formation of well-defined subgrain structures, as can be seen from Fig. 1.(c). Dislocations overcome dislocation barriers and continue to move, resulting in no well defined dislocation structure (see Fig. 1.(c)). Dipoles are observed, but no dislocation walls. We conclude that when the hole strength is very high, holes can disrupt the normal cell structure formed by dislocation interaction forces in the absence of holes. However, we expect that when the hole strength is relatively low, cells are more likely to form. A similar argument can be carried out to explain the effect of hole density on dislocation structure at fixed biaxial stress. Increases in hole density have the effect of increasing hole strength. At the lowest hole density, no dislocation rearrangement has occurred except very close to the origin where dipoles are seen, but no walls. As the hole density increases, the stress fields of holes help dislocations to rearrange themselves into low energy structures, therefore formation of dislocation cells is much more pronounced. However as the hole density continues to increase, the presence of many closely-spaced holes prevents the formation of a subgrain structure. This leads to separation of dislocations from holes, as can be seen from the distribution function (see in Fig. 2.(c)). Formation of dipoles is observed, but no walls.

5. ACKNOWLEDGEMENTS

The supercomputer (Cray X-MP/24) time of this research by the Alabama Supercomputer Center is gratefully acknowledged.

6. REFERENCES

Cottrell A H 1953 *Dislocations and Plastic Flow in Crystals* (London: Oxford University Press) 98

Gould H and Tobochnik J 1988 *An Introduction to Computer Simulation Methods Application to Physical Systems*, (New York:Addison-Wesley)

Gulluoglu A N 1991 Ph.D Dissertation, University of Alabama at Birmingham, Birmingham, AL, U.S.A.

Hirth J P and Lothe J 1968 *Theory of Dislocations* (New York: McGraw-Hill) 194

Maddin R and Cottrell A H 1955 *Phil. Mag.* **46** 735

McClintock F A and Argon A S 1965 *Mechanical Behavior of Materials* (Reading: Addison-Wesley)

Peach M and Koehler J S 1950 *Phys. Rev.* **80** 436

Weast R C 1975 *Handbook of Chemistry and Physics* (Cleveland: Chemical- Rubber)

Dislocation cell diameter during creep

J H Gittus, N M Ghoniem[*] and R Amodeo[**],

British Nuclear Forum, London, England
*University of California Los Angeles, California, USA
**Xerad Limited, Los Angeles, California, USA

ABSTRACT: A comparison is made between two different theoretical predictions of the dislocation cell diameter for alpha iron undergoing creep. One prediction derives from hypothesis that steady-state creep is a condition of thermodynamic equilibrium characterized by a minimum in the free energy of the specimen. The other prediction is obtained from a computer model of a crystal containing several hundred dislocations in which no such assumptions are made. Good agreement is obtained between the values of K, the ratio of cell diameter to mean dislocation spacing. Similar agreement is obtained for the constant K' expressing the inverse relationship between cell size and applied stress.

1. INTRODUCTION

When crystalline materials are subjected to stress at elevated temperatures the slow, time-dependent deformation process known as creep occurs. Often, during this process, the crystal dislocations form into walls, of small but finite thickness, outlining cells which are smaller than, but in some ways similar to, the grains. One of us (Gittus 1977) has developed an equation linking the strain-rate with a parameter, K, which is defined as the dimensionless ratio of the dislocation-cell diameter to the distance between dislocations, imagining the dislocations to be "smeared out" so that instead of being concentrated in the cell walls they uniformly fill all the space within the crystal.

It is the purpose of the work reported in this paper to test the validity of the method that had been adopted to arrive at the theoretical value of K. The procedure adopted is numerical analysis, using a Cray computer to model the movement of some hundreds of dislocations, each subjecting all the others to stresses, during simulated creep under an external stress in alpha iron.

2. ANALYTICAL MODEL AND THE VALUE OF K

In the analytical model of creep it is posited that a dislocation cell structure has formed and that steady-state creep is now occurring under an externally applied stress. The stress needed to produce a given creep rate is calculated, taking account of dislocation friction, climb, glide, dislocation sources, dislocation immobilization, and processes in the cell walls. The stress turns out to be a function of the self- diffusion coefficient D_v, Burgers vector b, shear modulus μ, temperature T, strain rate ε_s, and the ratio K, of cell diameter to mean dislocation spacing.

To derive K, for the low-stress limit, the free energy of the specimen is first calculated. It comprises the elastic strain energy plus the anelastic strain energy due to reversible bowing of the dislocations plus the line energy of the dislocations. The free energy is then differentiated with regard to K and the differential set equal to zero. From this a value of K is derived. Differentiation a second time reveals that this K-value corresponds to a minimum in the free energy and we have made the hypothesis that steady-state creep *is* a condition of thermodynamic equilibrium and therefore of minimum free energy.

The presumed state of thermodynamic equilibrium leads to the expression:

$$(\delta F)_{T,V} \geqslant 0$$

We use the lower case delta δ to indicate that the free energy is unchanged or increased for any small (finite) displacement of the system. The value of K which satisfies this condition has been derived (Gittus, 1977) and is given in Table 1. For alpha iron at temperatures where creep occurs, c_j, the thermal-jog concentration will generally lie in the range 0.01 to 0.001 whilst measured ratios of relaxed-to unrelaxed modulus generally indicate a value of order 0.005 for C' (it cannot exceed 0.05). The theoretical value of K therefore lies roughly in the range 5 to 15.

Table 1.   Theoretical relationship between the values of K
              and those of c_j and K' for various values of C'

C'	K	c_j	c_j (approx)	K'
0	1	3.61	3.16	1.65
0	5	5.50×10^{-2}	5.63×10^{-2}	3.36
0	13	5.11×10^{-3}	5.13×10^{-3}	7.30
0	20	1.75×10^{-3}	1.74×10^{-3}	10.80
0.001	10	9.42×10^{-3}	9.04×10^{-3}	5.44
0.005	1	3.60	9.12	1.65
0.005	5	5.20×10^{-2}	7.24×10^{-2}	3.11
0.005	13	3.80×10^{-3}	4.17×10^{-3}	4.48
0.005	20	1.25×10^{-3}	1.15×10^{-3}	6.60
0.05	1	3.49	5.13	1.61
0.05	5	3.87×10^{-2}	4.07×10^{-2}	2.17
0.05	13	2.37×10^{-3}	2.34×10^{-3}	2.36
0.05	20	6.87×10^{-4}	6.46×10^{-4}	2.51

2.1 The Constant K'

Table 1 also gives theoretical values for a second constant, K'. This is a second dimensionless constant, in this case linking the cell size L with the applied stress according to the relation

$$\sigma = k'\mu b/L$$

3. NUMERICAL MODEL

The numerical model of a crystal used in this work is that of Amodeo and Ghoniem (1988). It is solved by a computer code, DisloDyn, which simulates the motion of dislocations projected onto a two-dimensional (2-D) plane. Each dislocation is described by a position vector and a velocity vector. Initially a random set of dislocations is introduced into the reaction space. They are all mobile and are allowed at first to experience a relaxation phase in which there is no applied stress and the dislocations move to equilibrium positions. The external stress is then applied and the simulation proceeds as follows:

1. Computation of short-range interactions, giving rise to dipoles, nodes, or annihilations;
2. Computation of long-range forces acting on each dislocation, by summing *all* the individual long-range forces from all the other dislocations in the system;
3. Calculation of effective stress acting on each dislocation. This includes both the *internal* stress from other dislocations and the external stress;
4. Evaluation of the *mobility* of each dislocation: if the net stress falls below the friction stress, then a dislocation is *immobile*;
5. Computation of the velocity of each dislocation from the net stress acting upon it; and
6. Computation of dislocation sources from position and velocity data.

Any dislocation that leaves through one side of the system is deposited on the opposite side and allowed to re-enter (i.e, as though there was a neighbouring, identical system or patch of dislocations).

3.1 Long-Range Forces on Dislocations

To compute the elastic potentials in isotropic elasticity we substitute the force balance ($F_i = \sigma_{ij,j}$) into Hooke's Law ($\sigma_{ij} = \lambda u_{k,k}\delta_{ij} + (u_{i,j} + u_{j,i})$) to obtain:

$$(\lambda + \mu)u_{j,ji} + \mu u_{i,jj} = F_i.$$

For parallel dislocations u_i is a function of x_1 and x_2 only. In plane strain, $u_3 = 0$. The field equations are satisfied by two complex functions $\phi(z)$ and $\psi(z)$, such that:

$$2\mu u(z) = \kappa\phi(z) - z\phi'(z) - \overline{\psi}(z),$$

$$\sigma_{11}(x_1,x_2) + \sigma_{22}(x_1,x_2) = 2[\phi'(z) + \overline{\phi}'(z)]$$

$$\sigma_{22}(x_1,x_2) - \sigma_{11}(x_1,x_2) + 2i\sigma_{12}(x_1,x_2) = 2[z\phi''(z) + \psi'(z)]$$

where $u(z) = u_1(x_1,x_2) + iu_2(x_1,x_2)$ and $\kappa = 3 - 4\nu$.

The equations are satisfied everywhere, except at regions containing dislocation cores.

If we define a third potential function, ω, such that

$$\omega(z)= z\phi'+\psi$$

the field equations become:

$$2\mu u= \kappa\phi - (z-\bar{z})\overline{\phi}'- \bar{\omega},$$

$$\sigma_{11}+\sigma_{22} = 2[\phi'(z) +\overline{\phi}'(z)],$$

$$\sigma_{22} +i\sigma_{12} = \phi'+\bar{\omega}'+ 2(z - \bar{z})\overline{\phi}''.$$

The potential functions for a single dislocation at position $z=\xi_i$ are given by :

$$\phi_i = \phi_{0i} +\phi_{1i}, \qquad \omega_i = \omega_{0i} +\omega_{1i},$$

where ϕ_0 and ω_0 are for an infinite medium, and ϕ_1 and ω_1 are additional terms to satisfy stress boundary conditions

$$\phi'_{0i}= 2A/(z-\xi_i), \quad \omega'_{0i} = [2\bar{A}/(z-\xi_i)] - [2A(\xi_i-\bar{\xi}_i)/(z-\xi_i)^2],$$

where $A= \mu b_e/[2\pi i(\kappa +1)]$ and $b_e = b_1+ib_2$, is the generalized Burgers vector.

The functions ϕ_{1i} and ω_{1i} are obtained by boundary integrals, to ensure equilibrium at the boundaries of the computational cell.

The stress tensor components are computed at dislocation j from all dislocations i at positions ξ_i. Hence

$$\phi'_j = \Sigma_i \phi'_i \text{ and } \omega'_j = \Sigma_i \omega'_i,$$

The stress tensor components are rotated into the coordinate system of dislocation j:

$$\begin{pmatrix} \sigma_{11}^* & \sigma_{12}^* \\ \sigma_{21}^* & \sigma_{22}^* \end{pmatrix} = \begin{pmatrix} \cos\theta & \sin\theta \\ -\sin\theta & \cos\theta \end{pmatrix} \begin{pmatrix} \sigma_{11} & \sigma_{12} \\ \sigma_{21} & \sigma_{22} \end{pmatrix} \begin{pmatrix} \cos\theta & -\sin\theta \\ \sin\theta & \cos\theta \end{pmatrix}$$

The Peach-Kohler Force is computed on dislocation j:

$$F_i = (b_e.\Sigma_i^*) \times \xi_i,$$

where Σ_i^* is the stress tensor at dislocation j and is the sense vector ξ_i.

F_j is decomposed into glide and climb components

$$F_j = \begin{pmatrix} F_g \\ F_c \\ F_\xi \end{pmatrix}_j.$$

Velocities are computed at j

$$v_j = \begin{pmatrix} v_g \\ v_c \\ v_\xi \end{pmatrix}_j.$$

Displacements at j

$$r_j = \Delta t . v_j.$$

Displacements are transformed back into the global system and new positions computed.

3.2 Results for Alpha-iron

In Table 2 are given the results obtained from the numerical model for alpha iron. The friction stress was 14 MPa with $\mu = 81.2 \times 10^3$ MPa and $b = 2.51 \times 10^{-10}$ at the chosen temperature of 600°C.

Table 2. Results given by the numerical model for alpha-iron at 600°C.

Stress (MPa)	Cell Size L (micron)	K	K'
100	2.2	6.4	11
120	1.3	6.4	7
150	1.1	5.1	8

It is observed that the value of K lies in the range which the analytical model would lead one to expect (5 to 15). The value of K' is on the high side but within the analytic model's predicted order of magnitude. Both K and K' are similar to experimentally-measured values, reported in the literature and summarised by Gittus (1977).

4. CONCLUDING REMARKS

The experimentally determined values of K generally lie in the same range as those deduced from the analytical and numerical approaches adopted in the present paper.   There is therefore support for these alternative theoretical approaches.   In particular, the analytic approach derives from an analysis which gives an algebraic relationship between steady-state creep rate, and stress and temperature.   That relationship receives additional support from the present work since the value of K deduced from the analytic creep equation is similar to that derived from the numerical model of a crystal containing dislocations.

As expected from experiment, as the stress is increased then, in the numerical model, the cells decrease in diameter.

There is agreement between the two models and with experimental data in the case of the constant K', which expresses the proportionality between dislocation cell-size and the reciprocal of the applied stress.

ACKNOWLEDGMENT

This work was undertaken whilst one of us (JHG) was a University of California Regents' Lecturer at UCLA.

REFERENCES

Amodeo, R and Ghoniem, N.M. 1988. Dislocation Dynamics, Parts I and II.
Gittus, J.H., 1977 Philos. Mag. 35 (2), pp 293-300
UCLA/Eng-8951 and 8952/PPG 1242 and 1243.

Part 2
Radiation Damage in Materials

Molecular dynamics in radiation damage

M.W. Finnis

Max-Planck-Institut für Metallforschung, Institut für Werkstoffwissenschaft, Seestrasse 92, D-7000 Stuttgart, Germany

ABSTRACT: The technique of molecular dynamics as an instrument for the theoretical study of radiation damage is briefly reviewed. Better interatomic potentials and recent strides in computing power have increased the relevance and importance of molecular dynamics, and several insights have been gained. A new treatment of energy losses to electrons and its inclusion in the MOLDY6 code are discussed.

1. INTRODUCTION

It is a great pleasure to pay tribute to Ron Bullough here on the occasion of his sixtieth birthday, and I hope the editors will allow this more personal Introduction than is usual in conference papers. At the same time as he was busy developing and applying the continuum rate-theory of radiation damage, Ron Bullough remained very aware of the underlying atomistic processes and recognised the value of molecular dynamics (MD) to radiation damage, besides the importance of developing realistic interatomic potentials for MD. His activity and encouragement in these areas has contributed very significantly to the progress which we have made. I will summarize in this paper the technical aspects of this progress, devoting most space to a new treatment of electronic energy losses.

MD simulations, in which the coupled equations of motion of many interacting particles are solved numerically, were introduced to radiation damage by Gibson, Goland, Milgram and Vineyard (1960). They used MD to study the energies and behaviour of focussed collision sequences and to estimate the damage threshhold. For the next twenty five years the field advanced mainly in the USA, reaching a significant stage with the simulations by Diaz de la Rubia, Averback, Benedek and King (1987) of 5keV cascades in Cu. These demonstrated cascade melting and resolidifying, and introduced a new idea about how vacancy clusters form: it appears that a molten region of lower density solidifies radially inwards and concentrates vacancies into a central cluster. A new standard in the field had been set by this work, even though only a simple pairwise force model (the Gibson II potential) was used.

To my knowledge MD was at first used at Harwell for the study of point defect migration in simple metals (Tsai, Bullough and Perrin 1970) and for the simulation of liquids (Schofield 1973), involving energies up to a few eV and up to one or two thousand atoms. Our first MD work on the processes of radiation damage was to study of the sputtering of gold atoms with energies ~10eV, relating to experiments in the 1MeV microscope (Cherns, Finnis and Matthews 1977). The step up in scale to over 10,000 atoms, in order to simulate cascade damage, was made with deLeeuw and Dixon´s (1985) new code for the Cray 1. They implemented flexible periodic boundary conditions and a link-cell method of neighbour list updating which required much less memory than the earlier code. The simulation of processes of higher energy demands the use of bigger boxes of atoms and more megaflops of computer power. Current simulations have moved up to a few keV and over 100,000 atoms since the acquisition of a Cray-2, and the speed of these computations is now some 500× faster per atom than was possible for Gibson *et al.* on their IBM 704. Even desk-top computers are now used to simulate the dynamics of ~40,000 atoms.

Other technical improvements followed in 1988 starting with the extension of the deLeeuw-Dixon code to use simple N-body potentials. Subsequently a faster algorithm (Heyes and Smith 1987) was introduced for neighbour list accounting which exploited the architecture of vector machines such as the Cray. A new algorithm was introduced to automatically adjust the time step used for integrating the equations of motion during a simulation, which typically has to start very short for the collision phase of a cascade. For the first time it became feasible to simulate a block of over a million atoms using an N-body interatomic potential. The resulting MOLDY6 code (Finnis 1988) has been further enhanced recently and is now in use for MD-radiation damage simulations at Harwell (Foreman, Phythian and English 1991) and Livermore (Diaz de la Rubia and Guinan 1990). The latter authors present first results for a 25keV cascade simulation in Cu using a computation cell of 500,000 atoms. This state-of-the-art calculation demonstrates the formation of two subcascades and an apparently stable vacancy-rich region.

2. INTERATOMIC FORCES

The bugbear of atomistic simulations of real systems was always the inadequacy of pairwise models of the interatomic forces in metals; hence the standard and sometimes embarassing question to the seller of every MD simulation : "How sensitive are your conclusions to the details of the potential used?". Progress was made with the introduction of so-called embedded atom potentials (Daw and Baskes 1984) and N-body potentials (Finnis and Sinclair 1984). The important advantage of these and subsequent models of a similar kind was that they gave realistic point defect formation energies and surface energies, quantities which are expected to influence the course of cascade collapse, without costing significantly more computation time than conventional pairwise potentials. Recent studies (Foreman, English and Phythian 1991) indicate that the adequacy of the potentials is no longer a major problem for low energies in simple metals, including the noble metals and nickel. For the bcc transition metals the potentials are less reliable because of the greater importance of bond-bending forces, and further research is required to establish the usefulness of existing models in radiation damage.

All potential models in the literature assume that the electrons respond adiabatically to the motion of the ions. This is a dubious assumption for the dynamics of cascades, which may cool significantly by electron-phonon collisions and electronic heat conduction Thus the main question about interatomic potentials has rather shifted to their validity for energetic (keV) collisions in which the electrons no longer respond adiabatically. The remainder of this paper describes how we have treated this problem.

3. ELECTRONIC ENERGY DISSIPATION

3.1 Theoretical Model

The MD studies referred to have shown that an energetic damage cascade, sometimes called a *thermal spike*, resulting from a single primary recoil of a few keV, is fully formed within about 1ps. It then enters a phase of cooling and energy dissipation, during which time the survival of point defects and their clusters is determined. Until recently, MD simulations took no account of the inelastic collision of electrons with ions. As a consequence, in such simulations heat is conducted purely by a lattice mechanism. It is well known however that the thermal conductivity of metals is dominated by the *electronic* contribution, which might therefore be important in the cooling phase of cascades. A simple calculation using the experimental value of thermal conductivity predicts a cascade will quench after several tens of fs, which is too short a time for vacancy diffusion by the conventional mechanism, although novel mechanisms such as liquid state transport are not excluded. If the timescale of energy transfer from ions to electrons is slow enough, however, then the conventional MD description is valid, in which the energy of ionic motion is conserved, and considerable atomic rearrangements, including vacancy clustering, might occur during the cooling phase. Flynn and Averback (1988, to be referred to as FA) discussed this question in a quantitative way, and we extended their ideas and implemented them in MD simulation (Finnis, Agnew and Foreman 1991, to be referred to as FAF), as I will describe here in a brief but physically motivated way.

The situation we have modelled is summarised by the following equations for the ion temperature T_i and the electron temperature T_e in a spherically symmetric cascade

$$\frac{\partial T_i}{\partial T} = \frac{1}{\tau_i} \cdot r^2 \nabla^2 T_i - \frac{1}{\tau_{cool}} \cdot (T_i - T_e)$$

$$\frac{\partial T_e}{\partial T} = \frac{1}{\tau_e} \cdot r^2 \nabla^2 T_e + \frac{1}{\tau_{heat}} \cdot (T_i - T_e) \ . \tag{1}$$

The picture here is that the ion and electron temperatures are not in equilibrium within the thermal spike, but evolve according to the above equations. Four timescales or coupling constants are involved in this evolution. The diffusion of heat through the ions in the absence of electron-phonon scattering is described by τ_i, which is given in terms of the heat capacity σ, the density ρ and the lattice thermal conductivity κ_i by

$$\tau_i = \sigma \rho r^2 / \kappa_i \tag{2}$$

just as in standard heat diffusion theory. At this point, in order to define the other coupling constants, we must introduce some physically based assumptions. We assume as in the standard theory of electrical conductivity that the mean free path λ of electrons between collisions with phonons is inversely proportional to the local ion temperature :

$$\lambda = r_0 T_0 / T_i \ . \tag{3}$$

This leads to a timescale for electronic heat conduction

$$\tau_e = \frac{3 T_i r^2}{v_f r_0 T_0} \tag{4}$$

in which v_f is the Fermi velocity of electrons, r_0 is the Wigner Seitz radius of the atoms and T_0 is a characteristic temperature which embodies the strength of the electron-phonon coupling, as defined by FA. A further assumption we make following FA is that on each electron-phonon collision a characteristic phonon energy $k_B T_D$ is exchanged with the lattice, where T_D is the Debye temperature. This leads us to a characteristic timescale for the heating of the electrons :

$$\tau_{heat} = \frac{\pi^2 r_0 T_0}{3 T_D v_f} \; . \tag{5}$$

Finally the timescale for the cooling of the ions is given by equating the heat lost by the ions to the heat gained by the electrons, which leads us to

$$\tau_{cool} = \frac{\sigma \rho}{\gamma_e T_e} \cdot \tau_{heat} \tag{6}$$

in which the prefactor represents the ratio of lattice to electron thermal capacities.

We notice that a stronger electron-phonon coupling (lower T_0) implies a longer electron diffusion time τ_e; in other words a strong electron-phonon coupling inhibits heat conduction by the electrons. At the same time a strong electron-phonon coupling increases the rate of uptake of heat by the electrons by lowering τ_{heat}. This is responsible for a significant difference in the behaviour of Cu and Ni. To give a feeling for the magnitudes involved, I have plotted in Fig.1 values of the four timescales as a function of cascade size r, using the parameters for Cu and Ni tabulated in FAF. The ion temperature for this purpose was got by dividing a cascade energy of 5keV over the ions within radius r.

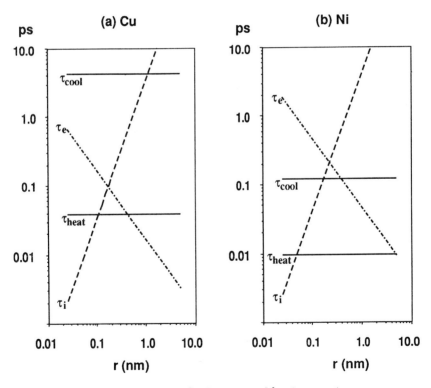

Fig.1. Timescales for the changes in electron and ion temperatures as a function of cascade radius.

We see that for all physically meaningful cascade sizes, that is for r greater than about 1nm, the electronic timescales are the shortest. Of particular interest is the crossing of τ_e and τ_{heat} which determines whether the electrons have time to heat up significantly themselves before they conduct away the heat they are receiving from the ions. There is a decrease in τ_e with r because as the ions cool, the mean free path of electrons become longer and they carry their acquired energy further between collisions.

For Cu this crossing of τ_e and τ_{heat} occurs already at r=0.4nm. In cascades of realistic sizes, the cooling rate characterised by electronic conduction is faster than the rate at which the electrons can acquire heat from the ions, so that their local temperature in a cascade will not be significantly higher than the ambient temperature. Considering the ion temperature in Cu, it falls fastest due to heat transfer to the electrons when τ_i becomes greater than τ_{cool}. This occurs already at 1.1nm, but the timescale is then 4ps, which is long on the scale of a single lattice vibration, so energy will be roughly conserved within the ionic system on that timescale.

The situation in Ni is different. The crossing of τ_e and τ_{heat} occurs at r = 5nm, so for realistic cascade sizes the electrons may also be heated significantly in the thermal spike. This is an effect of the much stronger electron-phonon coupling in Ni, which in turn is due to its partially full d-band of electrons. The timescale for ion cooling is also very short, namely 0.12ps, which is of the same order as the typical period of a high frequency lattice vibration. Hence we expect electronic cooling to be much more significant in Ni than in Cu. This may explain why vacancy loop production is observed to be more efficient in Cu than in Ni (Vetrano, Robertson, Averback and Kirk 1990) and why the fraction of vacancies contained within loops is higher in Cu (Smalinskas, Robertson, Averback and Kirk 1990).

3.2 MD Implementation

To implement the above physical model in the MD code required some more simplifications. At least for copper it is safe to assume that the electrons remain at room temperature. The alternative would be to introduce a spacially varying electron temperature into the MD code on some discrete mesh. We have not pursued this possibility further, although it does not seem difficult in principle. The approach we adopted in FAF was to associate T_i with the kinetic energy of an individual ion according to

$$T_i = m v_i^2 / 3 k_B \quad . \tag{7}$$

Each ion is made to suffer a damping force proportional to its velocity :

$$F_i = -\mu \cdot v_i \tag{8}$$

where the damping coefficient μ must be chosen to be consistent with the equations above. We impose this consistency by equating the rate of loss of energy from the ions to the electrons, namely $F_i v_i$, with the rate of decrease of the local ion energy :

$$-\mu \cdot v_i \cdot v_i = -3 k_B \frac{\partial T_i}{\partial t} \quad . \tag{9}$$

Hence substituting the ion cooling rate from (1) we get :

$$\mu = m \tau_{cool}^{-1} \cdot \left(\frac{T_i - T_e}{T_i} \right) \quad . \tag{10}$$

There is an unphysical divergence in this expression whenever an ion comes to rest, since we have defined the ion temperature in a local way, whereas the proper thermodynamic definition involves an average over a region of time and space large enough for fluctuations to be negligible. We have simply avoided the divergence in an empirical way by writing

$$\mu = m\tau_{cool}^{-1} \cdot \frac{T_i - T_e}{\left(T_i^2 + T_e^2 / 400\right)^{1/2}} .$$

(11)

We verified that the extra term in the denominator has no significant effect on the lattice vibrations in a room temperature simulation on a timscale of 10ps. We also made a MD run with an initial high value of the ion temperature and verified that it falls towards T_e with a time constant τ_{cool}, as required by equation (1). Preliminary calculations for a 500eV cascade (FAF) have shown that the stronger electron-phonon coupling of Ni inhibits the survival of defects, but in order to draw quantitative conclusions further work is needed on higher energy cascades.

4. CONCLUSIONS

I have presented an overview of recent developments in the technique of molecular dynamics in the simulation of radiation damage, with specific reference to the present MOLDY6 code. A new method of simulating electronic energy losses is described, which is based on electron-phonon scattering theory simply extrapolated to high energies. There are other approaches in the literature which lead to similar numbers for Cu, although their physical basis is the quite different theory of electronic stopping (Kaganov, Lifshitz and Tanatarov 1957, Caro and Victoria 1989).

REFERENCES
Caro A and Victoria M 1989 *Phys. Rev.* A **40** 2287
Cherns D, Finnis M W and Matthews M D 1977 *Phil. Mag.* **35** 693
deLeeuw S W and Dixon M 1985 *Phil. Mag.* A **52** 279
Daw M S and Baskes M 1984 *Phys. Rev.* B **29** 6443
Diaz de la Rubia T and Guinan M W 1990 *J. Nucl. Mater.* **174** 151
Diaz de la Rubia T, Averback R, Benedek R and King W E 1987 *Phys. Rev. Lett.* **59** 1930
Finnis M W 1988 Harwell Report AERE R13182
Finnis M W, Agnew P and Foreman A J E 1991 *Phys. Rev.* B **44** 567
Finnis M W and Sinclair J E 1984 *Phil. Mag.* A **50** 45.
Flynn C P and Averback R S 1988 *Phys. Rev.* B **38** 7118
Foreman A J E, English C A and Phythian W J 1991 AEA Technology Report AEA-TRS-2028, to be published
Foreman A J E, Phythian W J and English C A 1991 AEA Technology Report AEA-TRS-2031, to be published
Gibson J B, Goland A N, Milgram M and Vineyard G H 1960 *Phys. Rev.* **120** 1229
Heyes D M and Smith W 1987 *Information Quarterly for Computer Simulation of Condensed Phases,* No. 26, SERC Daresbury Laboratory, Daresbury, Warrington WA4 4AD, England, 68
Kaganov M I, Lifshitz I M and Tantarov L V 1957 *Sov. Phys. JETP* **4** 173
Schofield P 1973 *Comput. Phys. Commun.* **5** 17
Tsai D H, Bullough R and Perrin R C 1970 *J. Phys* C **3** 2022

Displacement cascades in metals

C.A. English, A.J.E. Foreman, W.J. Phythian*, D.J. Bacon** and M.L. Jenkins***

*   Radiation Damage Department., Reactor Services, AEA Technology, Harwell
     Laboratory, Didcot, Oxon, OX11 ORA, U.K
** Dept. of Materials Science and Engineering, University of Liverpool, P.O. Box 147,
     Liverpool, L69 3BX
*** Dept. of Metallurgy and Science of Materials, University of Oxford, Parks Road,
     Oxford, U.K.

ABSTRACT

In this paper we summarise the insight gained into selected features of cascade development which influence the subsequent microstructural evolution. The approach adopted combines experimental studies of vacancy loop formation in cascade cores with molecular dynamics modelling of cascade development using the MOLDY code. The latter has focused on simulating cascades in copper (and alpha-iron) with energies between 60eV and 10keV. Significant insight has been obtained from these simulations into the mechanisms separating vacancies and interstitials, the development of a thermal spike at the cascade site, the energy dependence of recoil production, and the clustering of interstitials on the cascade periphery. The emphasis in the discussion of vacancy loop formation is the impact of a high density of vacancy loops on microstructural development.

1. INTRODUCTION

The energetic recoils created in fast neutron-lattice atom collisions create collision casades in which the recoil, which may have 10's of keV in energy, loses its energy in a region a few tens of nm in diameter by creating a locally high concentration of displaced atoms. This paper is concerned with the understanding of the point defect motion, recombination and clustering that occurs in the cascade region as the high level of deposited energy dissipates into the surrounding lattice. The purpose is to summarise the insight gained into selected features of cascade development which influence the subsequent microstructural evolution. In previous review papers, English and Jenkins 1987, and English Foreman, Phythian, Bacon and Jenkins 1991, we reviewed progress in understanding a wider range of aspects of cascade development.

2. GENERAL EVOLUTION OF THE CASCADE.

Through following the general evolution of the cascade it is possible to identify the important point defect processes that will influence the subsequent microstural development.   Considerable insight has come from our molecular dynamics (MD) simulations of copper.   These MD studies, which used improved many-body potentials first derived by Finnis and Sinclair 1984, and Ackland et al, 1989, have been described in greater detail in Foreman et al 1991a,b, and Calder and Bacon 1991.   The simulations of low energy cascades (< 10keV) follow the generally accepted model of cascade development.   During the displacement phase (< 0.2 p sec) the initial primary atom loses its energy by a series of closely spaced collisions.   A heavily damaged region containing a large number of displaced atoms is established, which expands rapidly.   Interstitials are ballistically ejected from the centre of the cascade to form an interstitial rich shell around a central depleted zone.   At peak disorder (~0.3-0.5 p sec) replacement collision sequences (RCS) are observed emerging from the highly disordered central region, resulting in the formation of interstitial atoms some distance from the cascade centre.   At this time there is general agreement that the centre of the cascade is very hot and the cascade enters a thermal spike phase of development (see for example Diaz del a Rubia, Averback and Hsieh 1989).   Subsequently, considerable recombination occurs and the number of displaced atoms decreases rapidly.   MD simulations of copper and alpha iron for energies from 500 eV to 2 keV show that the number of displaced atoms increases with energy, and the time taken to achieve peak disorder also steadily increases, for example in copper it increases from ~0.2 psec at 500 eV to ~0.3 psec at 2 keV.   An interesting feature in alpha iron is that the form of the curves are similiar to that found in copper but the decay is far more rapid.   At 2keV the time taken for the number of displaced atoms to fall to half of the peak value is 0.75 psec in copper and 0.5 psec in alpha-iron.   The final configuration is generally a vacancy rich region surrounded by interstitial atoms (which may be clustered), as envisaged in the classical pictures of cascades, Seeger 1958, 1962.

It is the processes occurring during the thermal spike that are most important to the subsequent microstructural development. The central core of the thermal spike region has been almost instantaneously raised to a very high temperature, and it is during the decay of this hot region that the considerable point defect motion occurs. Indeed, the core of the cascade may be considered as a 'quasi-liquid' containing some cavitation. Vacancies may cluster in the centre of the depleted region (see below) and, as discussed below, the interstitials ballistically ejected may form clusters on the periphery of the depleted zone.

3. ENERGY DEPENDENCE OF DEFECT PRODUCTION.

3.1 Defect production in a perfect crystal.

There is general agreement that the recombination occurring during the thermal spike is greater than that allowed for in the accepted formalism for calculating damage due to Norgett, Torrens and Robinson (see for example Averback et al 1975, and Guinan and Kinney 1981). We have employed the MD simulations to follow the energy dependence of defect production at both 100K and 600K. Figure 1 shows the NRT efficiency factor k, as defined by Norgett et al, 1975, versus the recoil energy E for copper. This was derived from the equation:

$$N_d = k \, E_{dam} / 2 \, E_d \qquad (1)$$

where N_d is the point defect production (Frenkel pairs), E_{dam} is the energy available for damage (at low energy $E_{dam} = E$ as the electron energy loss is negligible) and E_d is the mean displacement energy. These results are based on over a hundred cascade simulations at each temperature with random knock-on directions and energies ranging from 60 eV (2 E_d) to 10 keV. At both temperatures there is a very sharp fall in k from the classical value of 0.8 at 60 eV to 0.4 at approximately 250 eV, which appears to be related to the transition from simple displacement events to a thermal spike. In contrast, from 250 eV to 2 keV there is a steady linear decrease in efficiency to 0.3 at 100K and 0.2 at 600K, which is associated with the increase in the size and lifetime of the thermal spike. The continued decrease in k occurs more slowly on going from 2keV to 10keV. We also include on Figure 1 values of k obtained from the alpha iron simulations of Calder and Bacon 1991 with E_d=30eV. Although based on a more limited data set (of 9 cascades), the efficiency factors are similiar to those of copper. This is somewhat suprising in view of the more rapid recovery of the damage in alpha iron.

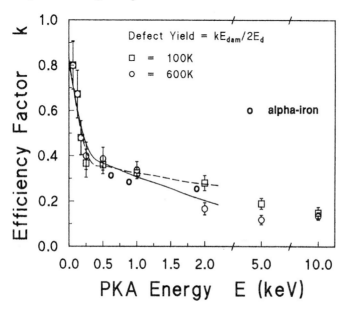

Figure 1   The damage efficiency factor, k, for defect production at 100 K and 600K in copper with recoil energies from 60 eV to 10 keV and random knock-on directions. Data from simulations of α-iron at 100K are also shown.

3.2. Effect of pre-existing features.

The data described above is for defect production in perfect crystal. It is interesting to examine how defect production is affected by the presence of pre-existing defect clusters. In the MD studies English et al 1990a 19 interstitial loop, a 19 vacancy loop, and a 19 vacancy void are set up in the block of atoms before bombardment with 1 keV atoms , which impact directly on the defect feature. In most of the cases studied, the net defect production was significantly lower than expected from similar simulations conducted on perfect crystals, typically only one extra Frenkel pair was created as against five expected in the perfect crystal. As this may be an important effect in microstructural evolution then it is necessary to determine the volume around the loop where defect production is reduced. The data in Figure 2 illustrates that the loop still influences defect production

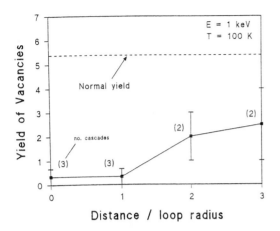

when the centre of the cascade is two or three loop radii from the centre of the loop. If this effect occurs at higher recoil energies then it suggests that this is potentially a significant effect in materials irradiated to high doses at low temperatures (<0.3 melting point). Here, a particularly dense micro-structure will build up and the MD simulations imply that defect production will be suppressed at least in low energy cascades. For example, if the microstructure consists of a high density of 2nm clusters, defect production will be suppressed in low energy cascades once the spacing approaches 6-8nm.

Figure 2   Plot of the yield of vacancies from 1keV cascades as a function of distance from a pre-existing 19 atom interstitial loop

4. POINT DEFECT CLUSTERING.

4.1 Defect clustering in a perfect crystal.

The general problem of point defect clustering in cascades has been pursued very actively in recent years, both experimentally and theoretically. In English and Jenkins, 1987 we reviewed progress in understanding vacancy loop formation from our experimental studies of fcc and bcc metals. Here we concentrate on the insight from recent MD simulations. In the hundred MD simulations conducted to date in copper, including energies of 5 keV and 10 keV, a fully formed planar vacancy cluster has been observed in only one 5 keV simulation. An outstanding feature of the present series of damage cascade simulations of copper at 100K and 600K has been the observation that the interstitial atoms to the periphery of the disordered zone during the displacement phase of the cascade frequently coalesce to form clusters.   This clustering arises because the strong elastic interaction between the stress fields of the self-interstitial atoms (SIA) enhances the probability that an individual SIA will combine with another of its kind rather than recombine with a vacancy. This becomes even more marked when several SIA have combined to form a 'loop' because the stress field strongly attracts other SIA's in the vicinity. Figure 3 shows a histogram of the number of atoms contained in clusters of various sizes for the computer simulations with PKA energies ranging from 500 eV to 10 keV. As the knock-on energy is increased the number of SIA produced in the peripheral region, where they can cluster, increases, so that at 2 keV and above an appreciable fraction of the SIA are being produced in cluster form. A sufficient number of cascades have now been simulated to provide some acceptable statistics for the production of these SIA clusters.   The data for simulations at energies of 10keV or less show the unexpected result that the average number of clusters N_{cl} containing more than two SIA's rises almost linearly with the PKA energy.

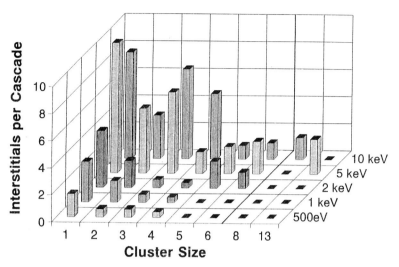

Figure 3   Histogram showing the mean number of interstitial atoms produced in clusters for knock-on energies from 500 eV to 10 keV.

It is to be emphasised that the ballistic ejection of interstitials is not sufficient to ensure continued separation of the vacancies and interstitials, as it is observed that single interstitials tend to move back into the cascade centre and recombine. It is the clustering of interstitials on the periphery that is the critical step in ensuring vacancy-interstitial separation. Further, these small clusters may be glissile and if they adopt a Burgers vector that is oriented towards the cascade centre then they are able to move towards the central depleted region and recombine with the vacancies in the centre of the cascade. However, it is observed that energy is released when the atoms combine and the resulting hot spot allows the atoms to find by experiment the Burgers vector that is most energetically favourable. It has been noted that in every case so far observed, the SIA clusters self-select a Burgers vector that is tangential to the cascade perimeter, which prevents them from moving into the cascade centre.

4.2 Influence of the clusters formed in cascades.

Here we illustrate the important influence the continuous creation of clusters can have on microstructural evolution. It has been recognised that vacancy loops will influence void swelling characteristics of fcc materials because vacancy loops are thermally stable up to 0.5 T_m, which is well above the 0.3-0.4 T_m normally observed for the onset of void formation, and will act as an alternative sink for the mobile vacancies that would otherwise go to voids. Stainless steel data suggest the fraction of vacancies may be tied up in small clusters, e, (see for example English and Jenkins 1987) is of the order of 0.01. At high doses there is no experimental evidence of the influence of the recoil sensitivity of vacancy loops production. The most thorough examination of the effect of vacancy loops, particularly on the correlation of neutron and simulation techniques, has come from employing a rate theory representation for the physical modelling of the microstructural evolution (Bullough et al, 1975, and Bullough and Quigley, 1983) to investigate the sensitivity of the swelling to the efficiency of vacancy loop production. The detailed

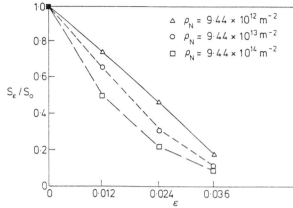

$\rho_N = 9.44 \times 10^{12} \, m^{-2}$
$\rho_N = 9.44 \times 10^{13} \, m^{-2}$
$\rho_N = 9.44 \times 10^{14} \, m^{-2}$

Figure 4   Variation of predicted swelling S_ε/S_0 at 100 dpa with dislocation density, ρ_N and ε in M316 stainless steel after neutron irradiation at 748K. Here S_ε/S_0 is the ratio of the swelling at a given ε to the value of swelling when ε is zero.

insight this provides has been set out in English, 1985 and only the conclusions will be provided here. The sensitivity of the predicted swelling at 475°C in neutron-irradiated M316 at 100 dpa to variations in both ε and dislocation density is examined in Figure 4. It can be seen that values of ε as low as .01 are sufficient to significantly reduce the swelling. Increasing ε reduces the predicted swelling, but most importantly, the degree of reduction is dependent on dislocation density, the greatest reduction is at the highest dislocation density.

5. SUMMARY.

The data summarised here have highlighted several aspects of cascade evolution which influence subsequent microstructural evolution. Namely, the detailed energy dependence of k, the effect of pre-existing features on defect production, significant interstitial clustering on the periphery of the disordered region, and the vacancy loop production. The studies reported also illustrate the benefits of using MD to investigate cascade processes. The implications of these phenomena to microstructural evolution are still being explored but the discussion of vacancy loop formation points to the important impact of the continuous creation of vacancy loops on dimensional changes at high doses.

5. REFERENCES

Bullough R., Eyre B.L., and Krishan K., Phil Trans Roy Soc. A346 81 1975

Bullough R., and Quigley T.M., 1983 J. Nucl Mater 44 318.

English C.A. J. Nucl Mater., 1985, 134, 71

English, C.A. and Jenkins, M.L.: Mater. Sci. Forum, 1987, 15-18, 1003.

Foreman, A.J.E., English, C.A. and Phythian, W.J.: 1991, AEA Technology Harwell Report, AEA-TRS-2028, submitted for publication.

Foreman, A.J.E., Phythian, W.J. and English, C.A.: 1991, AEA Technology Harwell Report, AEA-TRS-2029, submitted for publication.

Diaz de la Rubia, T., Averback, R. and Hsieh, H.: 1989, J. Mater. Res., 4, 579.

Norgett, M.J., Robinson, M.T. and Torrens, I.M., 1975, Nucl. Eng. Design, 33, 50.

Diaz de la Rubia, T. Averback, R.S., Benedek, R., and King, W.E.: 1987, Phys. Rev. Lett, 59, 1930.

Ackland, G.J., Finnis, M.W., Tichy, G., and Vitek, V.: 1989, Phil. Mag., 56, 735.

Finnis, M.W. and Sinclair, J.E.: Phil. Mag. A., 1984, 50, 45.

Calder, A.F. and Bacon, D.J.: to be published.

Seeger, A.: 1958, Proc. 2nd Int. Conf. Peaceful Uses of Atomic Energy, Geneva 1958, Vol. 6, UN, New York, p.250.

Seeger, A.: 1962, Radiation Damage in Solids, Vol. 1, IAEA, Vienna, p.101.

Measurement of replacement collision sequences in metals

T J Bullough

Department of Materials Science and Engineering, University of Liverpool, P.O.Box 147, Liverpool L69 0BX.

ABSTRACT: In order to identify the mechanisms by which interstitials are removed from displacement cascades, a series of low-energy heavy-ion irradiations at 4.2K of copper and molybdenum to very high doses have been used to produce and retain both near-surface vacancy and/or deeper interstitial components of damage in visible dislocation loops. The experiments provide evidence for a mechanism separating vacancies and interstitials around cascades only by relatively short distances of less than ~5nm. The results are contrasted with other experimental measurements and recent molecular dynamics simulations of displacement cascades and replacement collision sequences .

1. INTRODUCTION

1.1 Modelling of replacement collision sequences and their production around cascades

It is well established that during the evolution of a displacement cascade in a metal, interstitials are efficiently separated from cascade centres by a mechanism which can leave the vacancy-rich centre to collapse to a dislocation loop. Molecular dynamics (MD) atomistic simulations which realistically follow the motion of displaced atoms have given strong evidence for replacement collision sequences (rcs), operating along close-packed rows around the periphery of displacement cascades, as being able to athermally produce Frenkel pairs which are stable against recombination during the thermal-spike phase of cascade evolution and hence effectively separate the interstitials from their parent cascades. However, more recently these simulations have indicated that other processes may be important, such as the formation of highly glissile interstitial clusters and loops which may subsequently glide away from the vacancy rich cascade centre even at low temperatures.

Rcs operating around cascades so as to separate point-defects were first proposed by Seeger (1958) who slightly modified the focusons of Silesbee (1957) by combining their efficient transfer of energy with transfer of mass along the same rows. From computer models and geometric considerations (Robinson 1981) it was clear that for the minimum loss of energy and the maximum focusing by atoms surrounding the close-packed rows, rcs would occur efficiently along only the closest packed lattice rows and only at relatively low displacement energies (≤ 100eV), ie around the periphery of cascade volumes. The early MD simulations of rcs (Gibson 1960) illustrated the efficiency with which rcs can propagate along close-packed rows, with energy losses of only a few eV (typically 0.5-5.0eV) per replacement. More recently, Diaz de la Rubia (1989) showed that rcs with an average length of 2.3nm could be produced around a 5keV cascade in a static Cu lattice.

More realistic interatomic potentials and more efficient MD source codes have enabled full low-energy (up to 25keV) cascades and isolated rcs to be studied with more confidence. MD simulations of complete cascades initiated by displacements of up to 5keV in Cu and Ni have been possible (Diaz de la Rubia et al 1987, 1991), clearly showing the local melting which occurs in cascade centres and the short rcs around the edges of cascades. Foreman et al (1991a, 1991b) followed the evolution of cascades of up to 2keV (and more recently to higher energies) at a variety of temperatures and found that the rcs around cascades had energy losses which increases almost linearly with temperature. Even at 0K around 2keV cascades rcs were very short, averaging only 6.1±2.5 replacements with a maximum observed length of only 11 replacements (~2.6nm). While these lengths are sufficient for rcs to provide an efficient mechanism for separating point defects around cascades, the numbers of rcs per cascade was very low (only an average of 7.2 for a 2keV cascade). Of particular surprise in the Foreman work was the observed clustering of interstitials, created not by rcs but in the highly disordered interstitial-rich regions towards the periphery of the cascade volume, which survive recombination as the recrystallisation front retreats. These interstitial clusters contained almost the same numbers of interstitials as were produced by rcs, and they tended to form into interstitial loops with Burgers vectors tangential to the cascade periphery such that they can move away from the cascade by thermal glide. At 25keV interstitial loops have also been seen to be 'punched-out' by the pressure wave produced ahead of the expanding cascade (Diaz de la Rubia et al 1991). Especially for higher energy cascades, at higher temperatures and in alloys where rcs defocusing will be significant the role of interstitial clusters may be particularly important.

Foreman et al (1991b) also reported that for 2keV MD simulated cascades the efficiency factor (as defined by Norgett et al 1975) of about 0.3 for production of mobile interstitials by rcs or in interstitial clusters was considerably less than the value of 0.8 adopted as the NRT standard. Even lower defect production efficiencies were predicted at 10-20keV.

1.2 Experimental measurements of rcs.

While MD simulations are clearly the only way the evolution of individual cascades can be studied, there are three types of irradiation experiments, outlined below, which have been used to obtain information about point-defect separation processes around cascades, in particular to give estimates of separation distances and the role of rcs in these processes.

The only experimental technique which is able to examine the arrangement of point-defects around an individual cascade is that of field-ion microscopy (FIM). FIM studies of interstitial distributions around cascades produced by 30keV Cr^+ and 18keV Au^+ ions at 18K in tungsten were interpreted as being due to rcs with mean minimum ranges of 16±12nm (Wei and Seidman 1981). Although this range was considered to be an overestimate due to the sampling proceedure and the small number of interstitials detected, the studies gave direct evidence for point-defect separation over quite large distances around cascades. In the context of recent MD simulations, at the irradiation temperature used, any small interstitial clusters produced may also have been lost from the cascade volume examined.

An indirect estimate of the range of rcs was obtained from changes in the electrical properties of ordered alloys by disordering rcs following irradiation at temperatures below 10K. Results of irradiations with electrons (Becker et al 1968), thermal neutrons (Kirk et al 1977), fast neutrons (Takamura and Okunda 1973) and light ions (Sakairi et al 1981) have all been interpreted to give an estimate of the number of replacements per displacement varying from up to 50 in Cu_3Au (Takamura and Okunda 1973) to only 2.2±2.2 in Ni_3Mn (Becker et al 1968). The size of TEM imaged disordered zones in Cu_3Au produced by room temperature 10keV Cu^+ ion irradiation (Jenkins and Wilkins 1976) were found to correspond to only about ten replacements per displacement.

The final irradiation technique for studying point-defect separation processes around cascades was the high-dose irradiations of orientated single crystals as used by Hertel (1979) and Diehl et al (1968). They irradiated copper and niobium with 5keV Ar^+ ions at room-temperature so as to inject sufficient interstitials into the bulk of the specimens from near-surface cascades populations evolve. The rapid decrease in vacancy PDRE with dose and evolution of an interstitial loop population at higher doses is clearly illustrated in figure 3. At both 5keV and 10keV loop production is sensitive to surface orientation with much higher vacancy PDREs and lower interstitial PDREs for <001> compared with <011>. This surface orientation effect, plus the presence of deeper (>10nm) vacancy loops in the low-dose irradiation (Fig 1a) and the absence of deeper interstitial loops following Xe^+ ion irradiation (Fig 1c) as compared with self-ion irradiation (Fig 1b) provided good evidence for the importance of incident ion channelling in determining PDREs and the form of the depth distributions.

Following 10keV self-ion irradiation of molybdenum (Fig 4), interstitial loop depth distributions peak at about 10nm with a greater fraction of deeper loops for <001> than <112>, again probably due to incident ion channelling. The higher PDREs in molybdenum compared to copper were most likely due to the loss of near-surface vacancies to the surface during irradiation, following cascade collapse and vacancy loop glide to the surface. The difference in PDREs between <011> and <112> correlates with the interstitials agglomerating and unfaulting at depths such that image forces do not influence the unfaulting reaction and that only those loops which unfault to non-edge orientations are lost. Interstitial loops were also produced in Mo by self-ion irradiations at energies below 1keV and at temperatures up to 78K.

In all cases after high-dose irradiations PDREs are very low, such that if an NRT point-defect production efficiency of 0.8 is assumed then only between 0.1 and 1% of point-defects produced in copper and between 0.5 and 1.5% of interstitials in molybdenum survive in visible loops. These low PDREs, combined with the effects of channelling,

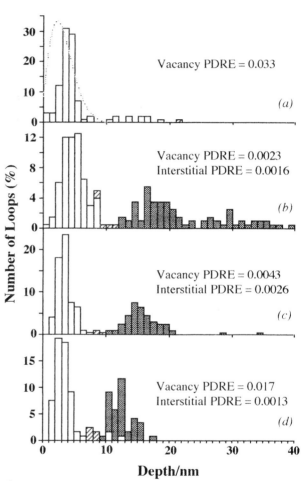

Figure 1. Loop depth distributions in copper following 10keV irradiation at 4.2K.
(a) 5.0×10^{12} Cu^+ ions cm^{-2} , <011> surface
(b) 1.0×10^{14} Cu^+ ions cm^{-2} , <011> surface
(c) 7.5×10^{13} Xe^+ ions cm^{-2} , <011> surface
(d) 5.0×10^{13} Cu^+ ions cm^{-2} , <001> surface
Key: ☐ vacancy loops
▨ interstitial loops
▨ uncharacterised loops
.... WSS energy deposition profile

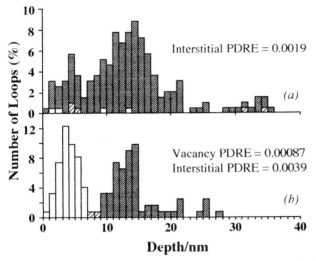

Figure 2. Loop depth distributions in copper following 5keV self-ion irradiation at 4.2K.
(a) 2.5x10^{14} ions cm^{-2}, <011> surface
(b) 2.0x10^{14} ions cm^{-2}, <001> surface
Key: as for figure 1

strongly indicated that while there must be a mechanism operating to separate the interstitials from cascade centres such that recombination is not occurring, it is not particularly efficient and probably operates only over relatively short (<5nm) distances. Long-range (>10nm) rcs ejecting interstitials from near-surface cascades would lead to much higher PDREs and would not have produced the different interstitial loop depth distributions observed following self-ion and xenon-ion irradiation (Figs 1b and 1c), which can only be accounted for by differences in incident ion channelling.

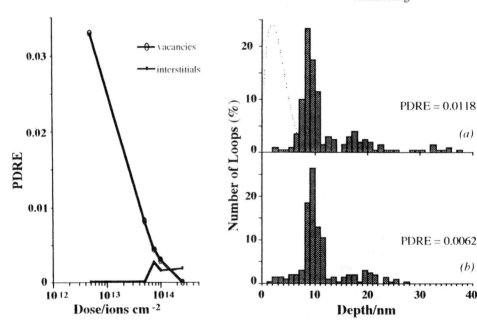

Figure 3. Vacancy and interstitial PDREs for 5-20keV <011> copper irradiations at 4.2K. The mean PDRE is used if more than one irradiation at a given dose.

Figure 4. Interstitial loop depth distributions in molybdenum following 10keV self-ion irradiation at 4.2K.
(a) 5x10^{13} ions cm^{-2}, <011> surface
(b) 5x10^{13} ions cm^{-2}, <112> surface
Key: as for figure 1

Clearly, the high irradiation doses result in considerable cascade overlap and high densities of damage, both of which may be acting to enhance the thermal (zero point motion) defocusing of rcs at 4.2K. Indeed, it is somewhat surprising that rcs could operate under these conditions sufficiently efficiently to separate out the vacancy and interstitial components of damage at all. While the ion energies are below the 25keV displacement energies for which cascade loop-punching was observed in MD simulations, the diffusion of small interstitial clusters away from 5 and 10keV cascades may be significant. Diffusion of such clusters was seen to occur in the MD simulations of Foreman et al (1991) at the 4.2K irradiation temperature used in these experimental studies. Further MD simulations of cascade evolution and a better understanding of low-energy incident-ion channelling are required.

Acknowledgements
I would like to thank Dr C A English and Dr B L Eyre for helpful discussions and suggesting and initiating the experimental work, which was undertaken partly in support of the Underlying Program of the UKAEA.

References
Becker D, Dworschak F, Lehmann C, Rie K T, Schuster H, Wollenberger H and Wurm J 1968 *Phys. Stat. Sol.* **30** 219
Bullough T J , English C A and Eyre B L 1991 *Proc. R. Soc. Lond A* (in press)
Bullough T J 1987 *PhD Thesis, University of Liverpool, UK*
Diaz de la Rubia T 1989 *PhD Thesis, State University of New York at Albany, Albany NY*
Diaz de la Rubia T, Averbeck R S, Benedek R and King W E 1987 *Phys. Rev. Lett.* **59** 1930
Diaz de la Rubia T and Guinan M W 1991 *Phys. Rev. Lett.* (in press)
Diehl J, Diepers H and Hertel B 1968 *Can. J. Phys.* **46** 647
Ecker K H 1974 *Rad. Effects* **23** 171
Foreman A J E, English C A and Phythian W J 1991a *Harwell Report AEA-TRS-2028*
Foreman A J E, Phythian W J and English C A 1991b *Harwell Report AEA-TRS-2031*
Gibson J B, Goland A N, Milgram M and Vineyard G H 1960 *Phys. Rev.* **120** 1229
Hertel B 1979 *Phil. Mag.* **A40** 313 and 331
Jenkins M L and Wilkins M 1976 *Phil. Mag.* **A34** 1155
Kirk M A, Blewitt T H and Scott T L 1977 *Phys. Rev.* **B15** 2914
Norgett M J, Robinson M T and Torrens I M 1975 *Nucl. Eng. Design* **33** 50
Reutov V F and Vagin S P 1984 *Sov. Phys. Solid State* **26** 633
Robinson M T 1981 *Topics in Applied Physics Vol. 47 (Sputtering by Particle Bombardment)* ed R Behrisch (Berlin: Springer-Verlag) 73
Sakairi H, Yagi E, Kujama A and Hasiguti R E 1981 *J. Phys. Soc. Japan* **50** 3023
Seeger A 1958 *Proc. 2nd Int. Conf. Peaceful Uses of Atomic Energy, Geneva* (New York: UN) **6** 250
Silsbee R H 1957 *J. Appl. Phys.* **28** 1246
Takamura S and Okunda S 1973 *Rad. Effects* **17** 151
Wei C Y and Seidman D N 1981 *Phil. Mag.* **A43** 1419
Winterbon K B, Sigmund P and Sanders J B 1970 *K. Dan. Vidensk. Mat. Fys. Medd.* **37** 14

Collision cascades and defect accumulation during irradiation

B.N. Singh[1] and C.H. Woo[2]
[1] Materials Department, Risø National Laboratory, DK-4000 Roskilde, Denmark
[2] AECL Research, Whiteshell Laboratory, Pinawa, Manitoba ROE 1LO, Canada

ABSTRACT: Considerations of the thermal stability of interstitial and vacancy clusters formed spontaneously within a cascade volume give rise to the concept of "Production Bias"; the production bias could be the dominant driving force for the void swelling. This concept has been used to calculate the void swelling in the transient as well as in the steady state regimes. The void swelling in the transient regime as well as its temperature dependence in the steady state calculated using this concept are found to be consistent with experimental results.

1. INTRODUCTION

Under cascade damage conditions vacancies and self-interstitials (SIA) are produced in high local concentrations and in a highly segregated fashion. This disposition leads to clustering of both vacancies and SIA's during the cooling down phase of the cascades (Woo and Singh 1991; Zinkle and Singh 1991). The problem of point-defect clustering in cascades was recognised by Bullough, Eyre and Krishan (1975) who included the consideration of vacancy loop formation within the cascades in the rate-theory treatment of void swelling. However, the effect of interstitial clustering was not included and the peak-swelling rate was found to be little affected by the vacancy clustering, so that dislocation bias is still employed as the dominant driving force for void swelling. Recently, Woo and Singh (1990, 1991) considered the effect of the clustering of both vacancies and interstitials and found significant effects on void swelling under cascade damage conditions at the peak swelling temperature. The temporal dependence of the production bias-driven void swelling in the transient regime has been investigated by Singh and Foreman (1991) and Singh, Woo and Foreman (1991). The concept of "production bias" was proposed as a dominant driving force for void swelling, the origin and physical basis of which are reviewed in the present paper.

2. CASCADE DAMAGE AND PRODUCTION BIAS

The classical picture of a cascade with a vacancy-rich core surrounded by SIA's at the periphery has been confirmed by computer simulation experiments using a binary approximation code (Heinisch 1990) as well as the molecular dynamics (MD) code (Diaz de la Rubia and Guinan, 1991, English, Phythian and Foreman 1990, Foreman, Phythian and English 1991). The MD simulations have demonstrated that a large fraction of vacancies and SIA's produced in a collision cascade annihilates via mutual recombination during the cooling-down phase of the cascade. A substantial fraction of the defects surviving the cascade quench forms clusters and only a small fraction escapes the cascade volume as freely migrating defects.

Numerous studies have reported evidence of vacancy clustering in cascades (see English and Jenkins 1987, English 1990, for reviews). As far as the clustering of SIA's is concerned, von Guerard and Peisl (1975) reported the evidence of SIA cluster formation directly in cascades in neutron irradiated copper already in 1975 (see also Grasse, von Guerard and Peisl 1984 and Rauch, Peisl, Schmalzbauer and Wallner 1989 for further results). The most direct evidence of SIA clustering in cascades has been provided by the MD simulations (Diaz de la Rubia and Guinan 1991, English et al. 1990, Foreman et al. 1991). Using a diffusion-based methodology, Woo, Singh and Heinisch (1990) have shown that a significant fraction of the SIA's produced at the periphery of a cascade may be immobilized in the form of clusters.

The vacancy clusters from which thermal emission of vacancies occurs can be considered as internal sources. At these temperatures, the SIA clusters are thermally stable (Ullmaier and Schilling 1980, Lam, Doan and Dagens 1985) and cannot provide mobile SIA's to the medium. Furthermore, the fraction of vacancies immobilized in the form of clusters is not likely to be the same as that of SIA's. It is then easy to see the inherent existence of a large asymmetry and hence a bias between the production efficiencies of mobile vacancies and interstitials under cascade damage conditions, i.e. a production bias, which is absent in the case of Frenkel pair production.

3. VOID SWELLING UNDER CASCADE DAMAGE CONDITION

Because of intracascade recombination during the thermal spike phase, the effective number of defects participating in subsequent microstructural evolution and defect accumulation is considerably smaller than calculated by the NRT model (Norgett, Robinson and Torrens 1975). In the present calculations, we therefore use an <u>effective point defect generation rate</u> G which is given by $G = (1 - \alpha)K$ where α is the fraction of defects that recombines in the cascade and K is the point defect production rate as calculated by the NRT model. Taking $\lambda_j (j = i,v)$ to be the fraction of defects that are immobilized by clustering, μ_i $(i = i,v)$ to be the fraction which escapes the cascade region and eventually annihilates at sinks, or recombines with free point defects of the opposite kind (after long-range migration) and $\varepsilon_j = \lambda_j/(1 - \alpha)$, we get (Woo and Singh 1991) $G = \mu_j K/(1 - \varepsilon_j)$ for $j = i,v$. Using this definition of G, it has been shown that the production bias is a potent driving force for the accumulation of vacancies in voids during cascade damage conditions (Woo and Singh 1991). In this analysis, it is implicitly assumed that the number density of the interstitial clusters is maintained at a steady-state value by some mechanism such as dislocation sweeping which is also the mechanism via which interstitial clusters are incorporated into the dislocation network.

We consider void swelling in a metal in which the dislocation structure is made up of a network of density ρ_N, vacancy loops of density N_{vL} and interstitial loops of density N_{iL}. We also denote, for the vacancies, the total dislocation sink strength by k_d^2, the void sink strength by k_c^2, and the total sink strength by k_v^2. Let p be the dislocation bias then we may write down the total dislocation sink strength for interstitials as $(1+p)k_d^2$, the corresponding total sink strength being k_i^2.

For the present steady-state point-defect concentrations, the rate equations for the vacancies and interstitials can be written as

$$(1-\varepsilon_i) \, G - \left[k_c^2 + (1 + p) \, k_d^2 \right] D_i C_i - \alpha_t C_i C_v = 0 \qquad \textbf{(1)}$$

$$(1 - \varepsilon_v) \, G + K^e - (k_c^2 + k_d^2) D_v C_v - \alpha_t C_i C_v = 0 \qquad \textbf{(2)}$$

where α_t is the coefficient of thermal recombination due to long-range migration, and K^e is the total vacancy emission strength for the various sinks, and has contributions from the voids, the interstitial loops, the vacancy loops, and the network. $D_v C_i$ are the diffusivity and concentration, respectively, of vacancies; $D_i C_i$ have similar meaning.

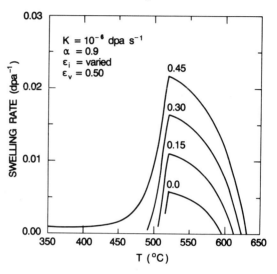

Figure 1

The calculated swelling rate as a function of irradiation temperature for various values of ε_i (= 0.0 - 0.45) at $K = 10^{-6}$ dpa s^{-1}, representing the case of fast reactor irradiation.

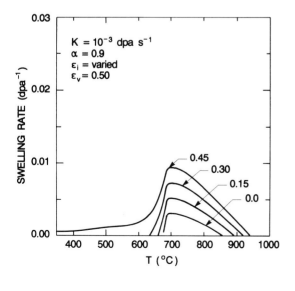

Figure 2

The same as in Figure 1 but at $K = 10^{-3}$ dpa s^{-1}, representing the case of heavy-ion irradiation

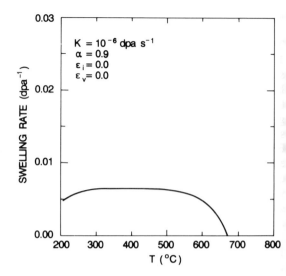

<u>Figure 3</u>

The same as Figure 1, but with ε_i and ε_v set to zero. Comparison with Figure 1 shows the effect of intracascade clustering of vacancies and SIA's.

Equations (1) and (2) can be solved straightforwardly and the solution is given in Woo and Singh (1991). Using experimentally derived quantities for the network dislocation density number densities of the voids and the interstitial loops (see Table 1, Woo and Singh, 1991) the swelling rate can be calculated. The variations of the calculated swelling rate with irradiation temperature are shown in Figures 1 - 4 for different values of dpa rate and clustering efficiencies ε_i and ε_v.

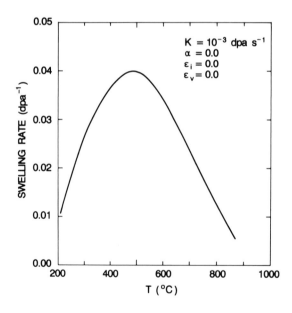

<u>Figure 4</u>

The same as in Figure 2, but for the case of low recoil energy irradiatin where the damage is produced homogeneously and in the form of Frenkel pairs instead of multi-displacement cascades,; this represents the case of HVEM irradiation.

4. DISCUSSION AND CONCLUSIONS

The results shown in Figures 1, 3 and 4 have been discussed in detail earlier (Woo and Singh 1991) and therefore only their main features will be described here. The swelling behaviour shown in Figure 2 should represent the case of heavy-ion irradiation with a significantly higher damage rate than that in the case of neutron irradiation (Figure 1). The comparison of the results shown in Figures 1 and 2 clearly demonstrates the effect of damage rate; for a given α, ε_i and ε_v the temperature of the onset of the high swelling rate as well as the peak swelling temperature are shifted towards higher temperature at the higher damage rate. Furthermore, the magnitude of the peak swelling is reduced at the higher dose rate. These effects arise due to enhanced recombination, and the low vacancy emission strength relative to the high production rate. The most important feature of the temperature dependence of the void swelling (Figures 1, 2) under cascade damage conditions is the sharp peak and very steep slopes on both the low and the high temperature sides of the peak; the steep slopes are characteristics of large activation energies (~ 3 eV) associated with both processes.

To demonstrate the effect of vacancy clustering in cascades alone, the results for $\varepsilon_i = \varepsilon_v = 0$ are shown in Figure 3. Comparison with Figure 1 shows that the vacancy clustering only affects the swelling rate at temperatures below the peak swelling. This effect originates from the accumulation of vacancy in vacancy loops.

It has been shown earlier (Woo and Singh 1991) that the same set of equations can be used to calculate and compare the case of neutron low dose rate, cascade damage and electron (high dose rate, Frenkel pair production) irradiations. The resulting swelling under electron irradiation is plotted against temperature in Figure 4. In contrast to the neutron damage (Figure 1) the temperature dependence of the swelling rate under electron irradiation conditions is much smoother on the low temperature side. This is because here the temperature dependence is governed by a recombination mechanism which has an activation energy of half of the vacancy migration energy (~ 0.7 eV) and is much smaller than that in the neutron case.

The general predictions of the temperature dependence of the steady-state swelling rate are in good agreement with experimental results on metals and alloys irradiated by fast-neutrons (Sloss and Davidson 1973, Edmonds et al. 1979, Brown and Fulton (private communication)). The low swelling rate was later confirmed at 400°C with steady-state swelling rate of 0.1% per dpa (Porter and Garner 1988).

It should be pointed out the dose dependence of the swelling in the transient regime has been calculated entirely in terms of the production bias (Singh and Foreman 1991, Singh, Woo and Foreman 1991). The calculated results are in a reasonable accord with the experimental results.

Since the strength of the production bias is determined by the efficiency of interstitial clustering, the swelling rate would be expected to be sensitive to material variables and radiation conditions, particularly the recoil energy. The production bias provides a natural explanation for the high steady-state swelling rate despite the very small fraction of free point defects.

ACKNOWLEDGEMENT

One of the authors (CHW) would like to express his gratitude to CANDU Owner's Group (COG) for funding this work, and to the Materials Department, Risø National Laboratory for hospitality during his stay at which this work was completed.

REFERENCES

BULLOUGH, R., EYRE, B.L., and KRISHAN, K., 1975, Proc. Roy. Soc., A346, 81.
DIAZ de la RUBIA, T. and GUINAN, M.W., 1991, Phys. Rev. Lett., in press; Int. Conf. on Physics of Irradiation Effects in Metals, May 20 - 24, Siófok, Hungary.
EDMONDS, E., SLOSS, W.M., BAGLEY, K.Q. and BATEY, W., 1979, Proc. Int. Conf. on Fast Breeder Reactor Fuel Performance, March 5 - 8, Monterey, pp. 54 - 63.
ENGLISH, C.A., and JENKINS, M.J., 1987, Mat. Sci. Forum, 1003.
ENGLISH, C.A., 1990, Rad. Eff. and Defects in Solids, 113, 29.
ENGLISH, C.A., PHYTHIAN, W.J., and FOREMAN, A.J.E., 1990, J. Nucl. Mater., 174, 135.
GRASSE, D., GUERARD, B.V., and PEISL, J., 1984, J. Nucl. Mater., 120, 304.
HEINISCH, H.L., 1990, Rad. Effect and Defects in Solids, 113, 53.
LAM, N.Q., DOAN, N.V., and DAGENS, L., 1985, J. Phys. F: Met. Phys. 15, 799.
NORGETT, M.J., ROBINSON, M.T., and TORRENS, I.M., 1975, Nucl. Eng. Des., 33, 50.
PORTER, D.L., and GARNER, F.A., 1988, J. Nucl. Mater., 159, 114.
RAUCH, R., PEISL, J., SCHMALZBAUER, A., and WALLNER, G., 1989, J. Nucl. Mater., 168, 101.
SINGH, B.N. and FOREMAN, A.J.E., 1991 (submitted to Philos. Mag.).
SINGH, B.N., WOO, C.H., and FOREMAN, A.J.E., 1991, Proc. Int. conf. on Physics of Irradiation Effects in Metals, May 20-24, Siófok, Hungary.
SLOSS, W.M. and DAVIDSON, A., Nov. 1 1973, TRG-Memo-Report-6300.
ULLMAIER, H., and SCHILLING, W., 1980, "Physics of Modern Materials", IAEA, Vienna, p. 301.
von GUERARD, B. and PEISL, J., 1975, J. Appl. Cryst., 8, 161.
WOO, C.H., and SINGH, B.N., 1990, Phys. Stat. Sol., B159, 609.
WOO, C.H., and SINGH, B.N., 1991, Philos. Mag., in press.
WOO, C.H., SINGH, B.N., and HEINISCH, H.L., 1990, J. Nucl. Mater., 174, 190.
ZINKLE, S.J., and SINGH, B.N., 1991, J. Nucl. Mater., in press.

Gas bubble nucleation in irradiated metals

S M Murphy[1] and M Fell[2]

[1]SD-Scicon UK Ltd, 1 Pembroke Broadway, Camberley, GU15 3XH
[2]Department of Mathematics, Bath University

ABSTRACT: The moment equations of Clement and Wood have been used as the basis for a simple model of the nucleation and growth of gas bubbles in an irradiated metal. The model predicts the number density of the bubbles, and the mean and standard deviation of their size distribution. The migration and coalescence of the bubbles is also included. The model is shown to give good agreement with experimental results.

1 INTRODUCTION

Gas bubbles form in irradiated metals as a result of the helium produced by transmutation reactions. The gas bubbles grow by the accumulation of gas atoms and irradiation-produced vacancies, and in many cases the bubbles reach a critical size where they no longer require more gas atoms to grow, but can grow simply by the accumulation of additional vacancies. Under these circumstances, the bubbles become voids, and tend to grow much more rapidly. This rapid growth causes the phenomenon known as void swelling. In addition, the presence of gas bubbles in the material can cause embrittlement of the metal.

This paper describes a simple model for the first stage of this process, namely the homogeneous nucleation of gas bubbles in an irradiated metal. The model is based on the moment equations suggested by Clement and Wood (1979) for describing the growth of a distribution of small clusters. Clement and Wood (1980) later extended their model to include both the growth and shrinkage of clusters under steady-state conditions. In this work, we have extended this model further to include a source term for the clusters and to include changes in the size distribution caused by coalescence of the clusters. This theory was then applied to the case of homogeneous nucleation of helium bubbles in metals during irradiation.

It is assumed here that the bubbles are in equilibrium, with the internal gas pressure balancing the surface tension of the bubble. Therefore the bubble radius is determined by the number of gas atoms in the bubble. This assumption will be valid during the early stages of irradiation, but large cavities will tend to be underpressured because they can absorb more vacancies than interstitials. This behaviour is not included in the present model, but can be described by the rate theory of void swelling developed by Brailsford and Bullough (1972).

2 FORMULATION OF THE MODEL

The present model includes single helium atoms, which can migrate through the material, and clusters of helium atoms containing two or more atoms. These gas atom clusters can grow by absorbing gas atoms, or shrink by losing gas atoms. Loss of gas atoms from clusters can occur because of thermal emission (which is likely for helium atoms only when the internal gas pressure in the cluster is extremely high) or because of interactions with the neutrons or charged particles causing the irradiation of the material.

The first step in developing the model is to obtain equations for the concentration of clusters containing n gas atoms, which is denoted by the symbol c_n. Throughout this paper, the concentrations are in fractional units, i.e. c_n represents the concentration of clusters containing n gas atoms per lattice site in the material. If migration and coalescence of clusters are neglected, the concentrations of these clusters of gas atoms are given by the equations

$$dc_1/dt = G - 2\beta_1 c_1 - \sum_{n=2}^{\infty} \beta_n c_n + \sum_{n=2}^{\infty} \alpha_n c_n, \tag{1}$$

$$dc_2/dt = \beta_1 c_1 - (\beta_2 + \alpha_2)c_2 + \alpha_3 c_3, \tag{2}$$

$$dc_n/dt = \beta_{n-1} c_{n-1} - (\beta_1 + \alpha_n)c_n + \alpha_{n+1} c_{n+1}, \tag{3}$$

where β_n is the probability that a cluster containing n atoms will absorb an additional atom, α_n is the probability that a cluster containing n atoms will emit an atom, and G is the production rate of helium atoms in the material. A more detailed discussion of this model, including expressions for the parameters α_n and β_n, is given in Fell and Murphy (1990).

It is impractical to solve the full set of equations (1) to (3) for the concentrations of the gas atom clusters. Instead, we consider only three parameters which can be used to characterise the bubble distribution, and obtain equations for these parameters. These three parameters are: the total concentration of clusters, denoted by N, the mean number of atoms in the clusters, $\overline{n}$, and the variance of the number of atoms in the clusters, M_2. These variables are defined by the relations

$$N = \sum_{n=2}^{\infty} c_n, \quad \overline{n} = \sum_{n=2}^{\infty} n c_n/N, \quad \text{and } M_2 = \sum_{n=2}^{\infty} (n - \overline{n})^2 c_n/N . \tag{4}$$

The equations for these quantities are

$$dN/dt = \beta_1 c_1 - \alpha_2 c_2, \tag{5}$$

$$\frac{d(\overline{n}N)}{dt} = 2\beta_1 c_1 - \alpha_2 c_2 + N(\beta(\overline{n}) - \alpha(\overline{n})) + \frac{M_2 N}{2}\left(\frac{d^2\beta(\overline{n})}{dn^2} - \frac{d^2\alpha(\overline{n})}{dn^2}\right), \tag{6}$$

$$\frac{d(M_2N)}{dt} = N(\beta(\overline{n}) + \alpha(\overline{n})) + \frac{M_2 N}{2}\left(\frac{d^2\beta(\overline{n})}{dn^2} + \frac{d^2\alpha(\overline{n})}{dn^2}\right) + 2M_2N\left(\frac{d\beta(\overline{n})}{dn} - \frac{d\alpha(\overline{n})}{dn}\right)$$

$$+ (\overline{n} - 2)^2\beta_1 c_1 - (\overline{n} - 1)^2\alpha_2 c_2, \tag{7}$$

where terms including moments of the size distribution of order 3 or higher are neglected. The concentration of single gas atoms is given by the equation

$$dc_1/dt = G - d(\overline{n}N)/dt. \tag{8}$$

Equations (5) to (8) form the basis of our model for gas nucleation. The concentrations of single gas atoms, c_1, and of clusters containing two gas atoms, c_2, appear in these equations. The value of c_1 is calculated explicitly, but the value of c_2 must be estimated. This is done by assuming that the size distribution of the gas atom clusters takes the form of a log-normal distribution. This distribution is illustrated in Figure 1. Using this assumption, the value of c_2 can be calculated from the values of N, $\overline{n}$ and M_2 (see Fell and Murphy, 1990 for further details).

Figure 1. Probability density function for the size distribution of gas bubbles. For this case the mean and standard deviation of the distribution are both 10.

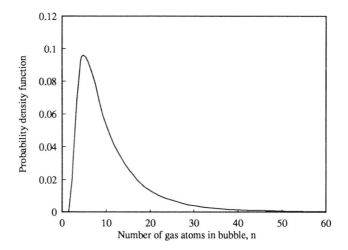

3 RESULTS OF MODEL

Calculations were performed using a wide range of input parameters, and the results of these are discussed in detail by Fell and Murphy (1990). In general, the average bubble size increases to a maximum, and additional gas atoms are accommodated by an increase in the density of these bubbles. This maximum bubble size depends on temperature, and is determined by a balance between the migration rate of gas atoms to bubbles and the rate at which the gas atoms are knocked out of the bubbles by the irradiation. This implies that a move from a fission reactor to a fusion reactor (which corresponds to an increase in the rate of production of gas without an increase in the rate of knock-on of gas atoms) will cause an increase in the cavity concentration, with very little change in the predicted cavity size. This behaviour also implies that the concentration of cavities increases indefinitely as the irradiation continues. By contrast, in real metals during irradiation, the concentration of cavities tends to saturate at moderate doses, whereas the cavity size continues to increase indefinitely.

There are two likely causes for this discrepancy. Firstly, the cavities can migrate and coalesce, causing an increase in the average cavity size and a reduction in the cavity concentration. Secondly, if the cavities become sufficiently large to grow as voids, then the average cavity size will increase significantly. In addition, the large cavities in the material will act as a strong sink for the gas, and will reduce the nucleation rate of additional gas bubbles.

The effects of bubble migration and coalescence on the bubble number density and on the bubble size are illustrated in Figures 2 and 3. The details of the model are given in Fell and Murphy (1990). Coalescence tends to increase the bubble size and reduce the bubble concentrations, especially at high temperatures. The strong temperature dependence seen in these figures arises because the bubbles are assumed to migrate by surface diffusion, and therefore the migration rate of the bubbles is strongly temperature dependent. These results suggest that the coalescence of gas bubbles is an important factor in determining the number of bubbles produced in an irradiated material, especially at high temperatures.

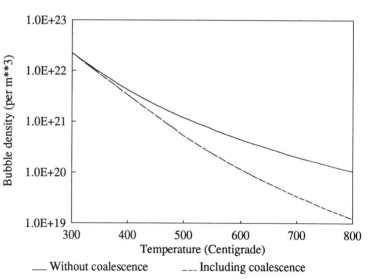

Figure 2. Effect of coalescence on the bubble number density. Parameters used are appropriate for Fast Reactor irradiation to 100 dpa.

___ Without coalescence   ___ Including coalescence

The results obtained from the model were compared with published results. As an example, Figure 4 shows a comparison of the results of the model with the cavity densities observed by Ayrault et al. (1981) in 316 stainless steel after dual beam irradiation. Three different gas production rates were used, with a irradiation damage rate of 3×10^{-3} dpa/s, and an irradiation temperature of 625°. Given the scatter in the data, the agreement between the experimental and theoretical results shown here is encouraging.

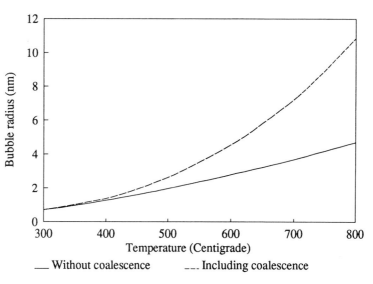

Figure 3. Effect of coalescence on the radius of a bubble containing the mean number of gas atoms. Parameters used are appropriate for Fast Reactor irradiation to 100 dpa.

___ Without coalescence   _ _ _ Including coalescence

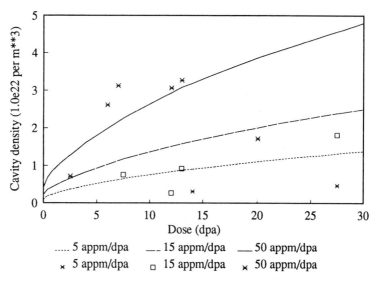

Figure 4. Comparison of calculated values of cavity density (shown by smooth curves) with experimental results of Ayrault et al. (1981) (shown by symbols).

..... 5 appm/dpa   _ _ 15 appm/dpa   ___ 50 appm/dpa
    × 5 appm/dpa   □ 15 appm/dpa   × 50 appm/dpa

The values of the cavity concentrations obtained from the calculations agree reasonably well with the experimental values for the 50 appm/dpa irradiation, but are in general higher than the experimental values for the lower gas injection rates. However, the experimental cavity concentration is likely to be an underestimate of the true concentration because of the presence of bubbles too small to be resolved in the microscope. Alternatively, the results may suggest that the calculations underestimate the role of coalescence, and this certainly appears to be the case for the 50 appm/dpa irradiation, where the measured cavity concentration falls at doses above 20 dpa. Ayrault et al.(1981) also found that the size distributions of the cavities were bimodal, which indicates that some of the bubbles have grown large enough to grow as voids, and this is also likely to influence the nucleation of further gas bubbles in the metal.

4 SUMMARY AND CONCLUSIONS

This paper describes a simple model for the nucleation and growth of gas bubbles in irradiated metals. This model is based on the moment equations first suggested by Clement and Wood (1979), which have been extended to include the creation of gas atoms within the material and the migration and coalescence of the bubbles. The calculations show that the coalescence of the bubbles can be a very important feature at high temperatures, with the cavity concentration being reduced significantly, while the cavity size is substantially increased.

The results of this model have been compared with experimental results. Given the scatter in the experimental data, the agreement between theoretical and experimental results is encouraging, but further work is needed to extend the model to larger cavity sizes where the cavities can no longer be described as equilibrium bubbles.

5 ACKNOWLEDGEMENTS

The authors wish to thank Dr. C.F. Clement of Harwell Laboratory for many helpful discussions during the course of this work. The work described in this paper was undertaken at Harwell as part of the Underlying Research Programme of the United Kingdom Atomic Energy Authority.

6 REFERENCES

Ayrault G, Hoff H A, Nolfi F V and Turner A P L 1981 J. Nucl. Mater. **103&104** 1035.
Brailsford A D and Bullough R 1972 J. Nucl. Mater. **44** 121.
Clement C F and Wood M H 1979 Proc. R. Soc. London **A368** 521.
Clement C F and Wood M H 1980 Proc. R. Soc. London **A371** 553.
Fell M and Murphy S M 1990 J. Nucl. Mater. **172** 1.

Ambient temperature precipitation of inert gases in metals

J H Evans

Materials and Chemistry Division, Reactor Services, AEA Technology,
Harwell Laboratory, Oxon., OX11 0RA, UK.

ABSTRACT: This paper describes some of the interesting phenomena that occur when metals are implanted with high concentrations of inert gases at ambient temperatures. These phenomena include helium-induced blister formation on implanted surfaces, fine scale precipitation of bubbles, bubble lattice formation, and the precipitation of the heavier inert gases in the solid phase. Helium desorption results are also discussed.

1. INTRODUCTION

As is well known, the behaviour of inert gases in metals is dominated by their high heat of solution leading, in most situations, to an essentially zero solubility and a strong tendency for precipitation as gas bubbles. Interest in the area has now been maintained over several decades with early studies on metals being inspired by their use as model systems for gas bubble behaviour in nuclear fuels, particularly UO_2, where the need existed to understand and predict the behaviour of krypton and xenon formed in the fission event. In this early work much of the physics of gas bubble evolution at high temperatures (eg. migration, coalescence, etc) was put in place. Generally speaking, the experimental studies at this time involved low concentrations of inert gas, usually helium up to about 100 ppm, inserted in the metal either by relatively high energy implants or via the reaction of boron-10 with thermal neutrons. High temperature annealing produced samples suitable for transmission electron microscopy (TEM) and for mechanical property testing.

Interest in higher inert gas concentrations, the main concern of this paper, developed in the mid-1970's initially as a result of the discovery of helium-induced blister formation on metals and its plausible relevance to fusion reactor technology where the first wall would be facing a D-T plasma with helium as one of its by-products. Considerable effort, making extensive use of scanning electron microscopy (SEM), was spent in studying blister formation following helium ion implants with energies in the range from a few keV up to about 1 MeV. This topic provides the start point for the phenomena discussed in this paper. There was a natural evolution from SEM to TEM to gain information on the helium implanted substructures; in turn, the discovery of high helium bubble densities led to investigations into fundamental aspects of bubble growth and mechanisms of blister formation. These aspects are discussed in the following sections, together with experiments into the growth of bubbles in the absence of displacement damage, bubble lattice formation and the more recent work on the heavier inert gases demonstrating 'solid bubble' formation in metals.

2. HELIUM-INDUCED BLISTER FORMATION

The extensive studies made on this phenomenon have shown that the injection of mono-energetic helium ions into all metals (and many other solids) at ambient temperatures leads, at a critical dose, to the formation of surface blisters. An extensive review has been written by Scherzer (1983). An example of helium-induced blister formation is seen in fig. 1(a). Although reasonably representative, even within this micrograph several variations in blister morphology occur, both in blister size and in lid appearance. Surface flaking can be another common variation but underlying mechanisms are thought to be identical. Among the parameters affecting blister formation are the helium energy, dose, sample temperature, and surface orientation. To a large extent these can be rationalised by the observation that blistering only occurs when a sub-surface helium profile is formed with a local peak of about 30 at.% helium. Thus, a monoenergetic peak is generally required though the energy range, from 1 keV up to at least 3 MeV, is very wide. At the lower end of this range, surface sputtering acts to prevent the build up of any critical sub-surface peak. Generally speaking, the sample temperature is unimportant provided the sample is below $0.5\ T_m$; above this temperature, the blister morphology is replaced by a pin-hole or sponge-like surface.

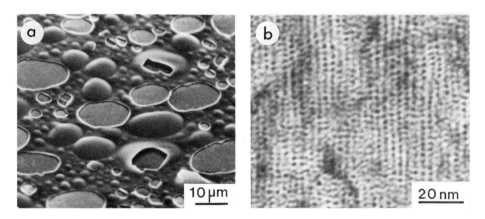

Fig.1 (a) Micrograph showing helium-induced blister on molybdenum; (b) transmission electron micrograph showing the underlying helium bubble substructure; in this example, bubble lattice formation is also seen.

The association of blister formation with a high concentration of insoluble gas naturally suggested that the blister phenomenon was gas driven. Nevertheless, in the late '70's there was considerable discussion on the possible role of lateral stress in the mechanism. As discussed in the next section, knowledge of the state of the helium in the metal was essential to put the physical mechanism of blister formation on a firmer footing.

2.1 Helium Bubble Substructures

The application of TEM to study the as-implanted substructures showed that for sample temperatures below about $0.4\ T_m$, all metals contained helium in high concentrations ($\sim 10^{25}/m^3$) of small bubbles, about 2 to 3 nm in diameter. A micrograph illustrating these bubbles is shown in fig.1b. An additional feature is the formation of a bubble lattice originally found on molybdenum, but later extended to other metals. This phenomenon, together with other cavity and precipitate lattices formed under radiation damage conditions, will be discussed in section 5.

Of relevance to the physics of bubble formation and growth was the observation that the bubble parameters were independent of formation temperature over a wide range which included temperatures below that of vacancy mobility. As examples, data for Mo and Zr have been given respectively by Mazey et al (1977) and Evans et al (1991). It was argued (Evans 1978) that helium bubble growth in this temperature range was independent of the displacement damage vacancies and grew instead by gas driven processes such as loop punching (Greenwood et al 1959). It might be added that when vacancies are mobile during implantation, the potential must exist for the well known bias-driven mechanism for cavity growth (Bullough and Perrin, 1972). However, this growth process is completely suppressed by the high bubble density dominating the sink substructure.

2.2 Interbubble Fracture

With the knowledge of the helium bubble substructure, the mechanism of blister formation soon evolved. If helium bubbles were overpressurised, then it became clear that their growth had to be limited by the local strength of the metal matrix between bubbles. Eventually the material between bubbles had to fracture - interbubble fracture - leading to gross sub-surface coalescence of bubbles to give a large plate-like bubble and finally surface blister formation. The general picture is shown in fig.2. The lack of blisters at high temperatures is explained by the availability of thermal vacancies and the resulting absence of any bubble over-pressure.

Fig. 2. Schematic of the processes involved in blister formation. The high densities of over-pressurised bubbles (a), leads to interbubble fracture (b & c) and the formation of a penny shaped crack (d) that eventually deforms the surface to give a blister.

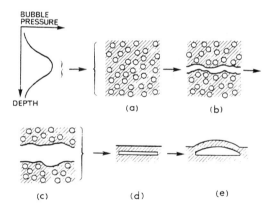

3. HELIUM DESORPTION SPECTROSCOPY

Helium desorption spectroscopy (HDS), in which gas release is measured during a linear temperature ramp, has provided unique information about bubble nucleation processes and the growth of helium clusters (van Veen 1989, 1991). The central feature in this work has been the implantation of helium at very low energies, 100 to 150 eV, to avoid any displacement damage, followed by a careful study of the size and temperature variations of desorption peaks with helium dose. In this way, it has been demonstrated that a single vacancy is an unsaturable trap for helium. Furthermore, in Mo and tungsten, up to 7 or 8 helium atoms can be bound to a vacancy, while additional helium leads to the formation of a Frenkel pair - allowing the helium trap to expand from a single vacancy to a divacancy. This is clearly the first stage of gas-driven cavity growth. As more helium is trapped at the growing helium cluster, the process continues inexorably although later growth, as shown by TEM (Evans et al 1983), involves the loop punching process. Surprisingly, the helium clusters at this later stage (containing 300 He atoms upwards) have a platelet morphology. It seems paradoxical that helium, the key atom in persuading vacancy clusters during irradiation to grow as voids rather than vacancy loops, itself precipitates two-dimensionally. However, this morphology is associated with the trapping of the self-interstitials expelled in the early stages of growth (suggesting the possibility of helium clusters acting as interstitial loop nuclei). Returning to the group of helium atoms bound to a single vacancy in Mo, HDS has demonstrated that a self-interstitial can recombine with the vacancy and release the helium back into the lattice. Although this and the growth processes described above have been studied using HDS under controlled conditions, clearly such information is essential in understanding helium and other inert gas behaviour in more practical environments.

4. HEAVY INERT GAS IMPLANTS

Although efforts were made, eg. with UVAS and EELS, to measure helium bubble pressures in implanted metals (see Donnelly 1985 for review), a significant advance in this area was made in the mid-80's. Two groups, Templier et al (1984) and vom Felde et al (1984), published electron diffraction data on aluminium showing that argon and xenon, respectively, were precipitating after ambient temperature implants in the solid phase. Evans and Mazey (1985, 1986) extended this result to krypton in a number of metals and it became clear that the precipitates, conveniently described as 'solid bubbles', existed in the solid phase by virtue of the high pressures that resulted from the growth processes. As in the helium case, the general picture was one of gas-driven bubble growth.

Fig. 3. An electron diffraction pattern of titanium implanted with krypton. The extra spots (arrowed), adjacent to the Ti (0002) reflections are from the krypton precipitates.

One fortunate aspect of the solid bubble result was that the precipitates were usually found to be epitaxially aligned with their close-packed plane parallel to the close-packed plane of the host metal. This behaviour was shown by Finnis (1987) to be energetically favourable. A further result, consistent with the known small energy difference between the fcc and hcp structures for the rare gas solids, was that in fcc metals the solid bubbles had the fcc structure expected for solid inert gas, but that in hcp metals their structure was hcp. The practical significance of the epitaxy was two-fold. Firstly, as shown in the fig.3 example, the presence of the solid bubbles was easy to detect via electron diffraction.

Secondly, the diffraction spots provided a semi-direct method of estimating bubble pressures by using the inert gas lattice parameters together with measured or theoretical equations of state. These approaches gave pressures accurate enough to show that bubble pressures drop during growth and that a good correlation exists between bubble pressures and the shear modulus of the host metal. Such a correlation is attractive to the model of gas bubble growth via loop punching – where the shear modulus is an important parameter. However, other correlations could exist and the exact mechanisms of gas-driven bubble growth under implantation are still a matter of discussion (Trinkaus 1991, Donnelly et al 1991). Nevertheless, it is clear that the solid bubble observations have made a useful contribution to the area. Additional experimental features of interest are the dark field imaging of the solid bubbles, Moire fringe observations, and heating experiments to follow precipitate melting (Evans and Mazey 1986). Further information is available in the recent review of Templier (1991).

5. CAVITY LATTICE FORMATION

As mentioned earlier, the bubble lattice first found for He bubbles in Mo (Sass and Eyre 1973) has been subsequently seen in many metals (see Johnson 1991 for recent review). Bubble lattices are also reported after heavier inert gas implants, eg. Kr in Zr (Evans et al 1989). All the observations appear to be a subset of the cavity lattice phenomena listed in table 1. Apart from their scale, the bubble and void lattice observations in metals are essentially identical.

Table 1. Summary of Cavity Lattice Observations

Material	Ordered defect (with examples of host materials)	Defect lattice structure
bcc metals	Voids (Mo,W,Nb,Ta,Fe) Helium bubbles (Mo,V,Fe) Neon bubbles (Mo)	bcc parallel to matrix
fcc metals and alloys	Voids (Ni,Al, stainless steel, Ni-Al,Ni-Cu alloys) Helium bubbles (Cu,Ni,Au, Al, stainless steel) Krypton precipitates (Ni)	fcc parallel to matrix
hcp metals	Voids (Mg) Helium bubbles (Ti) Krypton precipitates (Ti,Zr)	ordering into 2-D layers parallel to basal plane
alkaline earth halides	Anion voids (CaF_2, SrF_2)	simple cubic lattice
Alumina	Voids	linear alignment, perpendicular to basal plane
Silicon	Silicon oxide precipitates	simple cubic lattice, major axes parallel to Si

All these lattices have the characteristic of being formed under displacement damage conditions; for example, they are not formed or improved during annealing and not seen in helium bubble populations formed during tritium decay, where no displacements are created. A second characteristic is that their structure is either identical to the host lattice (eg. fcc in an fcc metal, bcc in a bcc metal), or closely related (eg. parallel to the basal plane in hcp metals). In addition, as the table shows, cavity lattices are also seen in non-metals. Many mechanisms have been proposed for cavity lattice formation; these range from the interaction models of Bullough and co-workers (Malén and Bullough 1970, Tewary and Bullough 1972) through to the more recent models applying possible anisotropies in the migration of self-interstitial atoms, either individually or as clusters. These mechanisms, together with a survey of experimental data, have recently been reviewed by the present author (Evans 1989).

Acknowledgement: The author thanks Ron Bullough for his encouragement and interest over the years in many aspects of the work described in this paper

REFERENCES

Bullough R and Perrin R C 1970 Int. Conf. on Voids Formed by Irradiation of Reactor Materials, Reading UK (Brit. Nucl. Energy Society) pp 79–107
Donnelly S E 1985 Radiat. Effects **90** 1
Donnelly S E, Mitchell D R G and van Veen A 1991 Proc. NATO Workshop on Fundamentals of Inert Gases in Solids, Bonas 1990 eds Donnelly S E and Evans J H (Plenum) pp 357–367
Evans J H 1978 J. Nucl. Mater. **76/77** 228
Evans J H, van Veen A and Caspers L M 1983 Rad. Effects **78** 105
Evans J H and Mazey D J 1985 J. Phys. F: Met. Phys. **15** L1
Evans J H and Mazey D J 1986 J. Nucl. Mater. **138** 176
Evans J H 1989 Proc. NATO Conf. on Patterns, Defects and Material Instabilities, 1989 eds Walgraef D and Ghoniem N M (Kluwer) pp 347–370
Evans J H, Foreman A J E and McElroy R J 1989 J. Nucl. Mater. **168** 340
Evans J H 1991 Proc. NATO Workshop on Fundamentals of Inert Gases in Solids, Bonas 1990 eds Donnelly S E and Evans J H (Plenum) pp 307–319
Finnis M 1987 Acta Met. **35** 2543
Greenwood G W, Foreman A J E and Rimmer D E 1959 J. Nucl. Mater. **4** 305
Johnson P B 1991 Proc. NATO Workshop on Fundamentals of Inert Gases in Solids, Bonas 1990 eds Donnelly S E and Evans J H (Plenum) pp 167–184
Malén K and Bullough R 1970 Int. Conf. on Voids Formed by Irradiation of Reactor Materials, Reading UK (Brit. Nucl. Energy Society) pp 109–119
Mazey D J, Eyre B L, Evans J H, McCracken G M and Erents S K 1977 J. Nucl. Mater. **64** 145
Sass S and Eyre B L 1973 Phil. Mag. **27** 1447
Scherzer B M U 1983 Topics in Applied Physics **52** 271
Templier C, Jaouen C, Rivière J-P, Delafond J and Grilhé J 1984 C R Acad. Sci., Paris **299** 613
Templier C 1991 Proc. NATO Workshop on Fundamentals of Inert Gases in Solids, Bonas 1990 eds Donnelly S E and Evans J H (Plenum) pp 117–132
Tewary V K and Bullough R 1972 J. Phys F. **2** L69
Trinkaus H 1991 Proc. NATO Workshop on Fundamentals of Inert Gases in Solids, Bonas 1990 eds Donnelly S E and Evans J H (Plenum) pp 369–383
van Veen A 1989 Mat. Sci. Forum **15–18** 3
van Veen A 1991 Proc. NATO Workshop on Fundamentals of Inert Gases in Solids, Bonas 1990 eds Donnelly S E and Evans J H (Plenum) pp 41–57
vom Felde A, Fink J, Muller-Heinzerling Th, Pfluger J, Scheerer B, Linker G and Kaletta D 1984 Phys. Rev. Letters **53** 922

Modelling gas bubble growth: theory and experiment

P J Goodhew

Department of Materials Science & Engineering, University of Liverpool

ABSTRACT: Experimental data from inert gas implantation and annealing experiments has often been used to draw conclusions about the operating growth mechanism. The problems associated with this approach are described, and some of its successes are highlighted.

1. INTRODUCTION

Gas bubble growth in solids is important for several reasons: The presence of gas bubbles can alter mechanical properties (eg embrittlement); bubbles can, on their own or by nucleating voids, result in a change of volume and density of a solid, and; gas bubble growth provides a fascinating and useful experimental tool for investigating physical behaviour (for example surface diffusion and thermal faceting). Two aspects of bubble behaviour are treated in more detail in other papers in this volume. Murphy (1991) discusses bubble nucleation while Evans (1991) discusses the precipitation of gas at low temperatures (a route to bubble nucleation) and Little (1991) concentrates on the effects of voids on swelling. In this paper the emphasis will be on the intermediate stages after inert gas has precipitated into a "bubble" but while its internal gas pressure is at or greater than the value needed to support the bubble. "Bubbles" in this context are thus either at equilibrium (with pressure $p = 2\gamma/r$ for a spherical cavity) or over-pressurised ($p > 2\gamma/r$).

The significance of the pressure inside the bubble is of course that, in the absence of external constraints, an overpressurized bubble has a driving force for growth, while an underpressurized cavity will tend to shrink. The key experimental fact which dominates our interest in radiation-related bubble behaviour is that inert gases are virtually insoluble in metals. This ensures that bubble populations are long-lived, and that they survive long enough for bubble migration (for example) to be studied.

2. GROWTH THEORIES

There are many ways in which a bubble can grow, and other mechanisms whereby, although some individual bubbles might shrink, the mean size of the bubble population grows. Most of the mechanisms which have been proposed are simple enough for the growth rate to be modelled in terms of a simple equation, at least if we assume that the cavity maintains a regular shape. Let us first consider, in principle, the mechanisms which can apply to a single bubble.

A cavity in a crystalline material can grow by collecting vacancies, by emitting interstitial atoms singly or co-operatively as a dislocation loop, or by migrating until it meets and coalesces with another bubble. The collection of vacancies is unsurprising, the punching of dislocation loops has been seen in the electron microscope (Greenwood et al, 1959) and plenty of indirect evidence has been found (eg Evans et al, 1983) for the emission of single interstitials from very small inert gas clusters (embryonic bubbles). Bubble migration is more surprising but there is now very clear evidence that bubbles in metals can migrate appreciable distances at temperatures as low as $0.55T_m$.

Barnes and Mazey (1963) showed in early TEM experiments at Harwell that small bubbles could migrate up a temperature gradient at appreciable rates, at temperatures only slightly above half the melting point. It was later shown at Surrey and Harwell that random "Brownian" motion also occurs in the absence of any long-range driving force (Tyler & Goodhew, 1980). The importance of bubble motion is that it leads to bubble collision and, since there is effectively no short-range repulsion between two bubbles, to coalescence. This inevitably leads to an increase in the mean bubble size but of greater significance is the fact that if vacancies can be supplied it will result in an increase in the total bubble volume (ie swelling).

The competing growth mechanisms - collection of vacancies, the transfer of vacancies and gas between bubbles (Ostwald ripening) and loop punching - may dominate at very high temperatures, or at very small bubble sizes and very high gas pressures, but for the regimes of most practical interest migration and coalescence is the most significant mechanism and has been studied in great detail.

For predictive purposes it would be useful to establish the rate-controlling mechanism for bubble migration. The transport of atoms from one side of the bubble to the other can take place by surface diffusion, by bulk diffusion or by transport through the vapour. These mechanisms lead to different growth rates and a great deal of effort has been expended on distinguishing between them. It certainly seems likely that surface diffusion will dominate at temperatures up to at least $0.6T_m$. It has now been established that this is the case for several metals, and by implication for most (eg Goodhew & Tyler, 1981).

The equations describing the rate of growth of bubbles of radius r by each mechanism are well known, and can be briefly summarised as:

Loop punching and interstitial emission: These mechanisms result in discrete athermal size changes and are not applicable to bubbles large enough to be measured individually.

Vacancy collection: $dr/dt = (DC/r)\{1-\exp[(2\gamma/r-p)a^3/kT)]\}$

Ostwald ripening: $dr_{mean}/dt = (D/r)\exp(-Q^{He}/kT)$

Migration and coalescence: $dr/dt = AD/r^n$

In the above equations A is a constant, n is a small integer which depends on the rate-limiting mechanism, D is the appropriate diffusion coefficient, C the equilibrium vacancy concentration and γ the surface energy.

It is tempting to conclude that the rate of bubble migration must be controlled by the rate at which the dominant diffusion mechanism can operate. However this is not necessarily the case: There may be an interface reaction which inhibits growth and becomes rate-controlling. A clear case occurs for large faceted bubbles (eg figure 1), when transport by surface diffusion is limited by the need to nucleate ledges on atomically-flat facets (eg Goodhew & Tyler, 1981). This effect proves to be useful in restricting the growth of bubble populations since large faceted bubbles become essentially immobile. The effect still operates when bubbles are very small, but the problem here is to determine (by TEM or otherwise) whether the bubbles are indeed faceted. It turns out that for most metals the size at which "ledge nucleation" becomes rate limiting is also the size at which it becomes relatively easy to see the facets in the TEM.

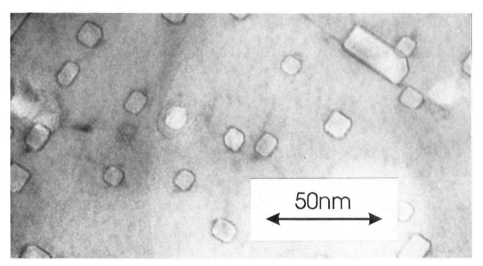

Fig. 1. Faceted helium bubbles in Nb-1%Zr annealed at 1250C for 8 hours.

Another effect which acts to slow bubble migration is the presence of extremely high gas pressures. A bubble containing gas at pressures high enough to reduce the "gas" phase atomic spacings to values similar to those in the solid will act essentially as a solid precipitate. Instead of surface diffusion around the inside of the bubble, the process in closer to interfacial diffusion at a phase boundary and thus is much slower. We have obtained experimental evidence that very small helium bubbles in nickel have been prevented from migrating until sufficient vacancies have arrived to lower the gas pressure (Marochov et al, 1987).

It is necessary to consider the extent to which the various growth mechanisms can operate in parallel and thus how the growth rates given by each equation should be summed. It is possible to show (Perryman & Goodhew, 1988) that in general the individual rates should simply be added, since the mechanisms are not mutually exclusive. For example migration and coalescence can occur while cavities are shrinking by emitting vacancies. This approach should be adopted when computing the total growth rate.

3. GROWTH EXPERIMENTS

Bubble growth experiments have largely consisted of inert gas implantation followed by annealing. In principle this approach offers a well-controlled experiment in which the temperature and inert gas content of the material are accurately known. If bubble size measurements can be made by electron microscopy, then a good comparison between theory and experiment should be possible. Bubble sizes between 2nm and 50nm should be accessible. However the comparison is not straightforward for a number of reasons which are considered and illustrated in the following paragraphs.

Size measurement: In order to determine bubble growth rates bubble size has to be measured at a series of time intervals. TEM of a series of specimens annealed for different times is the only practicable way of doing this. It is necessary to record the bubble images only very specific imaging conditions and to measure the appropriate fringe spacings (eg Cochrane & Goodhew, 1983). An error of 0.5nm in the diameter of a 5nm bubble represents a 10% uncertainty in diameter and a 33% uncertainty in volume. Since bubble populations invariable contain a range of sizes enough bubbles must be measured to enable the distribution to be specified. A single measure (eg mean size or mean volume) must be chosen to represent the population, since no growth theory apart from Ostwald ripening can deal simply with a distribution of bubble sizes. In much reported work the "bubble size" measure is not specified.

Bubble density measurement: This is in principle easier than bubble size measurement but is still fraught with difficulty. Central to density determination is the measurement of the foil thickness in the TEM, and the assumption that the density is uniform through the whole thickness of the foil. The former is difficult because most foils vary rapidly in thickness over the imaged area, while the latter is often untrue because of the presence of the foil surfaces and the inhomogeneous deposition of gas in many implantation experiments.

Composition: In most attempts to model bubble growth, material parameters must be incorporated. Data such as diffusion coefficients and surface energies are crucial when attempts are made to match absolute values of growth rates rather than slopes. However while the bulk composition of the material is often known, the composition at the region of interest (eg the bubble surface) may not be the same, and indeed may change during the experiment (eg Tyler & Goodhew, 1978). Implantation may also lead to the injection of impurities (Cox et al, 1989). In addition it is often not clear whether bulk values (eg of surface energy) are applicable to the internal surfaces of bubbles, which may be highly curved, strained or faceted, and are covered by a gas at high pressure.

Slopes: The most straightforward way to extract information about the growth mechanism would appear to be to plot ln(size) against ln(time). This has been done by many workers and an example is shown in figure 2. In principle the slope should indicate the growth mechanism but it is in practice difficult to obtain data with sufficient precision to permit confident discrimination between slopes of 0.2 and 0.25, for instance. In addition in some cases two or more mechanisms give the same slope. Trinkaus (1983) has suggested that it would be better in some circumstances to investigate the gas dose dependence of bubble size, but there are as yet few data available.

Solubility: The Ostwald ripening mechanism can only operate if gas can be transferred from bubble to bubble. The rate at which this can occur depends on the product of the gas solubility and its diffusivity. This product contains two poorly known parameters, one very small and one usually very large (at least for helium).

Relevance to reactor conditions: "Real" reactor conditions involve displacement damage and gas generation, probably pulsed. Both these effects are hard to simulate accurately, although it is possible to model gas generation (Singh & Trinkaus, 1991) and to estimate that displacement damage is unlikely to be significant (Luklinska & Goodhew, 1985).

Addition of mechanisms: Most of the mechanisms outlined here can operate simultaneously and it is therefore important to consider how they might combine. It seems most probable that the growth rates due to vacancy collection, Ostwald ripening and migration and coalescence can simply be added (Perryman & Goodhew, 1988).

Gas law: The gas law which is chosen to represent the behaviour of the atoms inside the bubble can radically affect the modelling of the growth rate, and Fichtner et al (1991) have shown that the ideal gas law is unlikely to be suitable for use with small bubbles (radius <50nm).

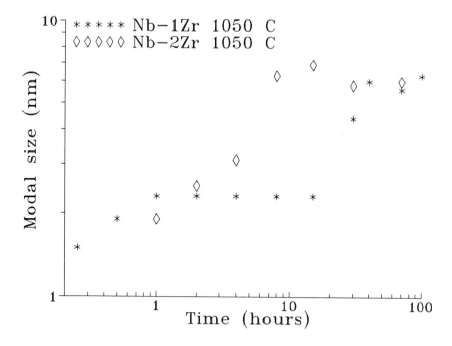

Fig. 2. Experimental data showing the growth of helium bubbles in two niobium alloys. The non-linearity of these data arises from the segregation of oxygen away from the bubbles during heat treatment (Goodhew & Tyler, 1981).

4. DISCUSSION

It would seem unlikely, from the foregoing discussion, that bubble growth experiments could be interpreted to give unambiguous information about growth mechanisms. However there have been some successes: Convincing evidence has been accumulated to show that surface-diffusion-limited migration and coalescence dominates bubble growth at low temperatures, while evidence is mounting that Ostwald ripening occurs, at least in fcc metals, at higher temperatures. Attempts have been made to construct mechanism maps which describe the regimes of temperature and bubble size and density in which each mechanism will dominate growth (Perryman & Goodhew, 1989). It must be borne in mind however that the experimental data extend over rather small fractions of the regimes of interest and that all the data are subject to the uncertainties outlined in this paper.

REFERENCES

Barnes R S & D J Mazey (1963) Proc Roy Soc **A275** 47-57

Cochrane B & P J Goodhew (1983) phys stat sol **A77** 269

Cox R J, P J Goodhew & J H Evans (1989) Nucl Instrum Meth **B42** 224-228

Evans J H (1991) This volume

Evans J H, A van Veen & L M Caspers (1983) Radiation Effects **78** 105-120

Fichtner P F P, H Schroeder & H Trinkaus (1991) Acta Met **39** 1845-1852

Goodhew P J & S K Tyler (1981) Proc Roy Soc **A377** 151

Greenwood G W, A J E Foreman & D E Rimmer (1959) J Nucl Mater **4** 305-324

Little E A (1991) This volume

Luklinska Z H & P J Goodhew (1985) J Nucl Mater **135** 201-205

Marochov N, L J Perryman & P J Goodhew (1987) J Nucl Mater **149** 296-301

Murphy S M (1991) This volume

Perryman L J & P J Goodhew (1988) Acta Metall **36** 2685-92

Perryman L J & P J Goodhew (1989) J Nucl Mater **165** 110-121

Singh B N & H Trinkaus (1991) Proc NATO ARW (Bonas) in press

Trinkaus H (1983) Rad Effects **78** 189-211

Tyler S K & P J Goodhew (1978) J Nucl Mater **74** 27-33

Tyler S K & P J Goodhew (1980) J Nucl Mater **92** 201-206

Void swelling resistance of ferritic steels: new perspectives

E.A. Little

Materials and Chemistry Division, Harwell Laboratory,
Oxfordshire. OX11 0RA

ABSTRACT: The principal features of the void swelling behaviour of bcc ferritic materials under irradiation are detailed and the response of technologically important higher alloyed steels compared with that of simple binary iron-chromium alloys and α-iron. The mechanisms and theories proposed to explain high swelling resistance are reviewed and re-evaluated in the light of current data trends.

1. INTRODUCTION

The formation and growth of voids in metals during high dose reactor irradiation can result in undesirable dimensional changes, leading to distortion of reactor core components. The confirmation of generic high swelling resistance in ferritic alloys is therefore of key technological significance. Thus selected commercial grades, such as the 10-12%Cr ferritic-martensitic steels – which possess the appropriate elevated temperature strength properties – are now specified for advanced fast reactor designs and are also strongly favoured for fusion reactor systems (Little 1987; Bagley et al 1988).

The boundaries of low swelling response and the mechanisms of swelling resistance in ferritics are topics which will underpin future alloy development and are considered in this paper. The subject of void swelling owes much to the pioneering theoretical studies of Dr. Ron Bullough FRS, who laid the foundations of the generally accepted rate theory approach, and then spearheaded subsequent refinements (e.g. see Bullough and Brailsford 1972; Bullough, Eyre and Krishan 1975; Bullough and Murphy 1985 and many other references in the literature). An understanding of swelling resistance in ferritics then followed naturally within this framework.

2. TYPICAL SWELLING TRENDS FOR FERRITICS

Ferritic materials studied under irradiation subdivide into four categories, viz. (i) α-iron and mild steels; (ii) binary Fe-Cr alloys; (iii) commercial non-transformable ferritic steels such as the 17%Cr grade; and (iv) commercial Cr-Mo bainitic (e.g. $2\frac{1}{2}$Cr1Mo) or ferritic-martensitic grades (e.g. 9-12%Cr with V, Nb or W additions – viz. EM10, HT9, FV448 or 1.4914 specifications as given in Bagley et al 1988). Their swelling response has been evaluated under neutron (fast reactor or mixed spectrum reactor), charged particle or electron irradiation conditions.

The swelling behaviour of ferritic and austenitic alloys are compared schematically in Figure 1, based on fast reactor irradiations in the temperature range 450°–550°C. All studies to date re-affirm that ferritic alloys form a general class of swelling resistant materials; however, distinct variations in response exist within the range of materials categories and irradiation conditions given above.

A recent data compilation (Little 1987) enables the following trends to be identified:

(1) Maximum swelling resistance occurs for fast reactor irradiations; higher swelling is observed under mixed-spectrum irradiations, as a consequence of greater helium production, and also under electron and ion irradiation conditions.

(2) Under neutron irradiation α-iron exhibits homogeneous void distributions with low void number density, with peak swelling at 400°–425°C and a low swelling rate of < 0.1% per dpa; a typical void distribution is illustrated in Figure 2. Evidence exists for a second swelling peak of lower magnitude centred at $\sim$ 510°C (Little and Stow 1979). Electron or ion irradiation shifts the peak swelling temperature upwards by 100° to 125°C.

(3) Additions of Cr to α-iron lead to progressive but modest reductions in swelling with a minimum at 3–5%Cr, but no change in peak swelling temperature under neutron irradiation (Little and Stow 1980; Gelles 1982). For Cr levels at or above 10% the situation is complicated by α' (Cr-rich ferrite) precipitation. The data demonstrate that Cr in solid solution does not account for the high swelling resistance of the commercial 12%Cr martensitic grades.

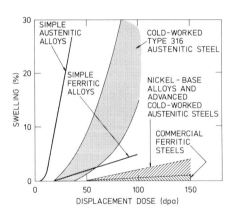

Fig. 1. Schematic comparison of fast reactor induced swelling in austenitic and ferritic alloys as a function of dose.

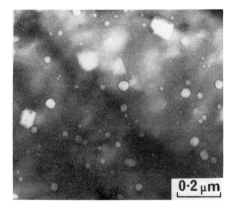

Fig. 2. Transmission electron micrograph showing a typical void distribution in α-iron after fast reactor irradiation to 23 dpa at 420°C.

(4) Recent fast reactor data on commercial 9-12%Cr ferritic-martensitic steels and 17%Cr fully ferritic steels have validated their swelling resistance to doses > 100 dpa (Bagley et al 1988; Gilbon et al 1989). Small levels of swelling of $\sim$ 0.5% detected at the highest doses appear at irradiation temperatures of 400°-450°C, i.e. corresponding to the peak swelling temperature for α-iron. This suggests that swelling suppression mechanisms act so as to extend the incubation dose for void formation and/or lower the void growth rate. Higher levels of swelling in 12%Cr steels ($\sim$ 0.1% per dpa) accompanied by homogeneous populations of small voids are observed in electron irradiations (Little 1985).

3. THE MECHANISM OF VOID SWELLING

The physical processes responsible for void nucleation and growth during elevated temperature ($\sim$ 0.3 - 0.5 T_m) irradiation have been enunciated by Brailsford and Bullough (1978). Vacancies and interstitials are lost either by mutual recombination or by absorption into sinks such as dislocations. Surviving interstitials aggregate into dislocation loops which expand, coalesce and finally form a dislocation network. Surviving vacancies cluster along with gas atoms (generally helium produced by transmutation) to form embryonic cavities. The dislocations act as biased sinks and preferentially absorb more interstitials than vacancies due to differing drift interactions. There is thus a net excess vacancy flux into neutral sinks such as the void embryos. When the latter contain a critical number of gas atoms (or, equivalently, reach a critical radius), biased-driven growth as voids takes place, leading to macroscopic swelling.

4. VOID SWELLING SUPPRESSION MECHANISMS

Several mechanisms which may limit void swelling in ferritic alloys have been advanced based on the above concepts, as follows:-

(1) Dislocation Loop Structure - A novel explanation proposed by Bullough et al (1981) is linked to the relative development of the two possible perfect interstitial dislocation loop geometries in the bcc lattice. Both species result from shear of a common a/2<110> faulted loop nucleus (Eyre and Bullough 1965) to give loops with b = a<100> (with low probability) or a lower energy loop with b = a/2<111>; this shear is predicted at small nuclei sizes of 16 atoms (Bullough and Perrin 1968). The dislocation reactions are:-

(i)   a/2<110> + a/2<110> → a<100>

(ii)   a/2<110> + a/2<001> → a/2<111>

The dislocation bias controlling swelling depends on the magnitude of the Burgers vector, |b|; the model thus postulates that a<100> loops act as strongly biased sinks relative to the more neutral a/2<111> loops (plus existing dislocation network), and the latter absorb the net excess of vacancies caused by growth of the a<100> loops, resulting in swelling suppression. Relatively few a<100> loops form initially, but shrinkage of a/2<111> loops by vacancy influx and renucleation events results ultimately in a predominantly a<100> loop population. The latter was observed in the early fast reactor irradiation data on FV448 steel (Bullough et al 1981), and is confirmed in recent irradiations to 45 dpa (Little et al 1991b) - see Figure 3.

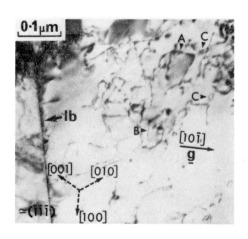

Fig. 3. Transmission electron micrograph showing dislocation loops near a lath boundary (lb) in 12%Cr martensitic steel after fast reactor irradiation to 45 dpa at 465°C. Loops A and B are rectilinear a<100> type; loops C are a/2<111> type. The <100> directions marked are projections in the (111) plane.

New key evidence supporting this mechanism is that of Muroga et al (1989) based on electron irradiations of a ferritic Fe-10Cr-1Ni alloy, who observed (a) that a<100> loops within a dual population grew at the expense of a/2<111> loops; and (b) nickel enrichment by non-equilibrium solute segregation occurred at a<100> loops but not at a/2<111> loops. These results are consistent with a higher bias for the a<100> loop geometry.

(2) Solute Interactions - Solute-point defect trapping can enhance recombination, leading to reduced vacancy supersaturations and in turn reduced void nucleation and growth rates (Hayns and Williams 1978; Little 1979). Carbon-vacancy binding energy estimates range from 0.41 to 0.85 eV (see references cited by Little 1987) and imply dominant interstitial element effects which can explain the low swelling of relatively pure α-iron with 50 ppm C+N (Little and Stow 1979). Substitutional elements may also be involved, and swelling in model ferritic Fe-10Cr-X alloys correlates with solute misfit strain for a range of solutes (Gelles 1989). Indirect evidence for strong point defect-silicon interactions is also available (Morgan et al 1990). Dominant solute-dislocation interactions (viz. Cottrell atmosphere formation) may also limit swelling by either reducing the dislocation bias or inhibiting dislocation climb (Little 1979).

(3) Low Intrinsic Bias - Calculations of Sniegowski and Wolfer (1984) indicate that the dislocation bias is lower in bcc iron compared to fcc iron due to smaller relaxation volumes for interstitials in the former; this would account for low void growth rates. However, Mansur and Lee (1991) now discount major differences in bias based on analysis of void growth rates.

(4) Microstructural Effects - Heterogeneous voidage associated with lath and subgrain boundary regions in martensitic steels imply that these features act as strong neutral sinks which retard swelling (Ayrault 1983; Maziasz et al 1986). Recent solute segregation studies indirectly confirm that these features are dominant sinks in these steels (Morgan et al

1990). Nevertheless swelling resistance is maintained in the 17%Cr fully ferritic steels in the absence of a subgrain structure. However, fine-scale precipitates of a suitable type (e.g. α' Cr-rich ferrite) may also act as efficient recombination centres that contribute to swelling resistance in these steels (Gelles and Thomas 1984).

(5) <u>Sink Competition Effects</u> – Mansur and Lee (1991) propose that low swelling microstructures result from domination by either dislocation or cavity sinks such that the controlling sink acts as a recombination centre. This model does not explain the large swelling differences between austenitic and martensitic steels, but can account for the modest variations in swelling rates in the various published data on the ferritics, which can be linked to the observed void densities.

5. GAS EFFECTS

The key role played by helium in stimulating void nucleation is confirmed by the higher levels of swelling ($\sim$ 0.5% at 47 dpa/400°C) observed in mixed-neutron-spectrum reactor irradiations in Ni-doped 9%Cr and 12%Cr steels (e.g. Maziasz and Klueh 1989) – promoted by the higher helium generation rates from the Ni (n,α) reaction – and in high gas input charged particle irradiations and/or triple beam studies using helium and hydrogen together (e.g. see Farrell and Lee 1987). High helium levels essentially switch the balance of bubble plus cavity distributions towards a larger voidage component, as modelled in current theories based on a critical bubble radius-to-void conversion during the incubation dose followed by bias-driven void growth (Odette 1988; Mansur and Lee 1991).

Recent ion irradiation data under high gas implantation conditions (600 ppm He) indicate significantly lower swelling in a 13%Cr ferritic oxide-dispersion strengthened (ODS) alloy containing a high density of small yttria particles compared with a standard 12%Cr martensitic steel

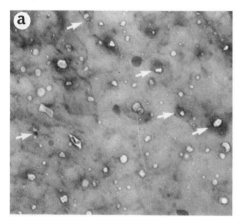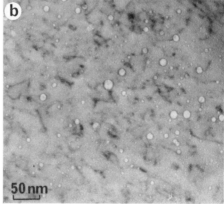

Fig. 4. Transmission electron micrographs of typical cavity distributions in (a) 13%Cr ferritic ODS alloy (swelling = 0.24%) and (b) 12%Cr martensitic steel (swelling = 0.49%) after ion irradiation to 50 dpa at 475°C with 600 appm helium. Examples of cavities associated with oxide particles in the ODS alloy are arrowed.

(Little et al 1991a). The void distributions are shown in Figure 4(a) and (b) and illustrate a significantly lower cavity concentration in the ODS alloy, with many cavities of irregular shape associated with oxide particles. The results imply that the oxide particles act as effective traps for helium, and may point the way to future ferritic alloy development for fusion reactor applications.

6. CONCLUDING REMARKS

The void swelling resistance of ferritic steels remains an area of high research and technological interest. Further experimental data are desirable under transient irradiation conditions (e.g. temperature cycling), high gas input conditions to high doses (> 100 dpa) and with varying sink distributions in order to explore the limits of behaviour, and in concert with theoretical modelling.

ACKNOWLEDGEMENTS

The studies described were undertaken as part of the Corporate Research Programme of the UKAEA.

REFERENCES

Ayrault G 1983 J. Nucl. Mater. 114 34.
Bagley K Q, Little E A, Levy V, Alamo A, Ehrlich K, Anderko K and Calza Bini A 1988 Nucl. Energy 27 295.
Brailsford A D and Bullough R 1978 J. Nucl. Mater. 69-70 434.
Bullough R and Brailsford A D 1972 J. Nucl. Mater. 44 121.
Bullough R, Eyre B L and Krishan K 1975 Proc. Roy. Soc. A346 81.
Bullough R and Murphy S M 1985 J. Nucl. Mater. 133-134 92.
Bullough R and Perrin R C 1968 Proc. Roy. Soc. A305 541.
Bullough R, Wood M H and Little E A 1981 ASTM Spec. Tech. Publ. 725 593.
Eyre B L and Bullough R 1965 Phil. Mag. 11 31.
Farrell K and Lee E H 1987 ASTM Spec. Tech. Publ. 955 498.
Gelles D S 1982 J. Nucl. Mater. 108-109 515.
Gelles D S 1989 ASTM Spec. Tech. Publ. 1046 73.
Gelles D S and Thomas L E 1984 Proc. Topical Conference on Ferritic Alloys for use in Nuclear Energy Technologies (USA:AIME) 559.
Gilbon D, Seran J L, Cauvin R, Fissolo A, Alamo A, Le Naour F and Levy V 1989 ASTM Spec. Tech. Publ. 1046 5.
Hayns M R and Williams T M 1978 J. Nucl. Mater. 74 151.
Little E A 1979 J. Nucl. Mater. 87 11.
Little E A 1985 Phys. Stat. Sol. (a) 87 441.
Little E A 1987 Materials for Nuclear Reactor Core Applications (London: BNES) 47.
Little E A, Mazey D J and Hanks W 1991a Scripta Metall. 25 1115.
Little E A, Morgan T J and Faulkner R G 1991b unpublished.
Little E A and Stow D A 1979 J. Nucl. Mater. 87 25.
Little E A and Stow D A 1980 Metal Sci. 14 89.
Mansur L K and Lee E H 1991 J. Nucl. Mater. 179-181 105.
Maziasz P J, Klueh R L and Vitek J M 1986 J. Nucl. Mater. 141-143 929.
Maziasz P J and Klueh R L 1989 ASTM Spec. Tech. Publ. 1046 35.
Morgan T J, Little E A and Faulkner R G 1990 ASTM Nashville Conf-in press.
Muroga T, Yamaguchi A and Yoshida N 1989 ASTM Spec. Tech. Publ. 1046 396.
Odette G R 1988 J. Nucl. Mater. 155-157 921.
Sniegowski J J and Wolfer W G 1984 Proc. Topical Conference on Ferritic Alloys for use in Nuclear Energy Technologies (USA:AIME) 579.

Modelling coolant channels for improving the performance of heavy water reactors

E. J. Savino

Comisión Nacional de Energía Atómica, Av. del Libertador 8250, 1429 Buenos Aires, Argentina

ABSTRACT: In Argentina, nuclear electricity is provided by means of two thermal natural Uranium heavy water reactors:Atucha I and Embalse. The design of each reactor solves the problem of separating coolant from the heavy water moderator using different concepts: Atucha I involves a pressure vessel within which coolant channels and moderator are located; Embalse is a CANDU 600 reactor with pressure tubes containing the fules and coolant flow. Both concepts use Zr alloy tubes, that are limiting factors in reactor life and performance. In this work we summarize our own efforts in studying both theoretically and experimentally dimensional changes either by growth or creep and chemical and structural stability. Performance and structural mechanics models for the whole channel are also developed in order to solve problems specific to these reactors. The latter are also summarized and actions under way in both stations are reviewed.

1. INTRODUCTION

The main purpose of this work is to summarize some applications of modelling radiation damage and mechanical behaviour to improve the performance of Argentinean nuclear power reactors. In Argentina two power reactors are integrated to the electric system: Atucha I, a Siemens design of 345 MWe, and Embalse, a CANDU reactor of 600 MWe. A third reactor, Atucha II, of 690 MWe is under construction. Both reactors burn natural Uranium fuel and are therefore cooled by heavy water. The coolant flow and moderator water must be separated and it is in this respect where the main difference in the design of either type of reactor appears. CANDU reactors include in their design horizontal pressure tubes at the so called fuel channels. These pressure tubes contain and support the fuel element bundles and are the main barrier between the high pressure water of the coolant and the moderator water. While at Atucha, coolant and moderator are at the same high pressure; therefore, the whole core is contained within a pressure vessel. The vertical coolant channels of Atucha serve as guides for the fuel elements and channels for the water flow.

The small neutron capture cross section of Zr and the good mechanical and corrosion properties of its alloys determine that both fuel cladding and inside components of the reactor core be constructed either of Zry or ZrNb (the latter for the pressure tube case). However, the mainly hcp structure of Zr alloys and its high chemical activity present two disadvantages with respect to cubic materials. First, the resulting mechanical properties are highly anisotropic and the material suffers small but noticeable shape change (growth) under irradiation. Secondly, it is highly sensitive to hydrogen pick up (Deuterium at reactor conditions). We shall see later how these facts may affect the life of the components.

program for coolant channel replacement in case it were needed. This program in turn involves a relatively intense research and development activity now under way.

3. INSURING THE PERFORMANCE OF FUEL AND COOLANT CHANNELS

3.1 CNE Fuel Channels and CNA-I Coolant Channels

Our early awareness of eventual problems after a prolonged PT-CT contact in the CNE decided us to start several actions towards insuring the PT end of life integrity. Three main lines are covered: 1) Surveillance; 2) Early Leak Detection; 3) GS Repositioning. The surveillance program includes the periodic and In Service Inspection (ISI) according to international standards plus locally-decided, more demanding, inspections integrated within a Research and Development "ad hoc" program. The second line, early leak detection, is a combined effort of engineers and operators to achieve the best conditions in this respect. The third point is an extended work. It includes not only the execution of single channels Separators Location and Repositioning (SLARing) during scheduled plant shut downs but also the feasibility of PT replacement. All this effort is based on a fluent contact with CANDU owners throughout the world, joint R and D with the CANDU Owners Group (COG), own R and D works (some of which are summarized in coming Sections) and establishing an as good as possible data base of the metallurgical properties of our own CNE tubes.

For the CNA-I coolant channels, we cannot benefit from the international experience in eventual problems in the same way as for a CANDU reactor. The main experiences are the designer's (SIEMENS), gained mainly through a prototype reactor (MZFR), and our own operating and repairing ones. An extended upgrading of the CNA-I will now be tried, being under study the eventual change of some channels during scheduled shut downs. This is encompassed within the R and D efforts mentioned below.

3.2 Research and Development in Zr alloys

We shall summarize our efforts in this field but, in accordance with the main subject of this meeting, emphasize theoretical and modelling work. Historically, our first efforts with Zr alloys were centered in mastering the process of fuel cladding and components fabrication. This is under way and 100% of the fuel burned at the CNE and CNA-I are manufactured at our factories. R and D is mainly concentrated in performance evaluation and its relation with fabrication variables. "In pool" and laboratory work is carried out and our own cladding and fuel performance modelling code (Harriague et al.) is being used. This code has the advantage of being based on sound material modelling. This approach is so successful that some ideas originally developed for simulating cladding performance by Savino and Harriague can be used for modelling reactor channels, as shown in the example of next section.

In relation to fuel channels, we are developing a data base of the metallurgical properties of the PT at the CNE; this contains analysis and measurements of specimens from the tube off cuts obtained during the mounting operation. SLARing equipment performance is monitored in the mock-up. Non destructive testing equipment is being developed and the laboratory work is intense. Experiments aim mainly to delayed hydrogen cracking (DHC) and corrosion studies. Modelling of the fuel channel structural behaviour is done through codes developed "ad hoc" by S. Terlisky. On the theoretical side, we have started, together with AECL groups

In this work we first summarize, in Section 2, standard problems in pressure tubes reported mainly in the Canadian literature. Our own experience on eventual problems in coolant channels for Atucha I (CNA-I) is also summarized there. In Section 3 we give a succint description of the joint approach taken at the Comisión Nacional de Energía Atómica (CNEA) by operators, engineering and research and development groups to insure the performance of those vital parts of our Central Nuclear Embalse (CNE) and CNA-I and II stations. Among those efforts, theoretical and experimental studies on microstructural properties and their relation with macrostructural mechanical performance are crucial to properties prediction and modelling. Our projects are therefore strongly based on basic metallurgy studies of Zr alloys. Some of these studies are also summarized in Section 3. Finally in 4, we show an example of an application of modelling that covers from point defects statics and dynamics calculations to the fitting of reactor measurements. Section 5 sums the Conclusions of the work.

2. FUEL AND COOLANT CHANNELS

CANDU pressure tubes: Recently, G.J.Field summarized the main problem caused by irradiation deformation in CANDU reactors. ZrNb corrosion and Deuterium uptake is revised by Urbanic et al. Two main groups of problems arose in the CANDU reactors due to these properties. One, dimensional problems due to larger axial elongation than that considered in the design of the first series of reactors, led or will lead to a somewhat large hardware work at Bruce and Pickering stations. Second, H/D pick up, together with a relatively large stress field, determines hydride or blister growth and eventual break of pressure tubes. Hydrides badly oriented due to high residual stresses appeared in Pickering and Bruce rolled joints, while a series of blisters
developed in the G16 tube of Zry2 in Pickering gave rise to a sudden rupture in 1983. Those blisters developed by an H/D migration to cooler regions of the tube, which in turn occurred because the garter springs (GS) separating Calandria Tube (CT) from Pressure Tube (PT) were misplaced. The CT are in thermal contact with the moderator, sensibly cooler than the coolant water, and they should be thermally isolated from the PT by a gas (CO_2) gap.

Pressure tubes at the CNE have enough length for allowing growth without any hardware work; however, care should be taken for the end of life conditions especially if power cycling is required by the electric network. Also, our channels are made of Zr 2.5%Nb which has a smaller H/D pick up rate than the Zry2 tubes. But, as in several CANDU reactors, some garter springs are outside their design position; therefore, PT-CT contact is possible as it has been experimentally verified through inspections in 1986 and 1989.

CNA-I coolant channels: At the CNA-I, in August 1988, a reactor shut down was performed due to a loss of reactivity; it was later realized that this loss was caused by the rupture of a coolant channel (R-06). In turn, it was found that a nearby lance guide tube (W3) had broken earlier being not only the origin of the rupture of R-06 but damaging several inside parts of the reactor as well. A major repair was undertaken and "on site" and metallurgical examinations of core materials were performed. The reactor was restarted in January 1990 and it has been operating since then. The repair and restart operations imposed the installation of systems and procedures for early damage detection. Metallographic studies allowed to discover that Zry oxidation and H/D pick-up was larger than expected although material ductility was still retained. This imposed including in the reactor backfitting a and under COG's auspices, to develop improved models of intergranular interaction for creep and growth simulation (C. Tomé). In the next Section, some new semi-empirical modelling for reproducing "in reactor" dimensional measurements for both fuel and coolant channels is discussed.

A similar amount of effort is also devoted for the CNA-I coolant channels. Non destructive techniques are being developed and new inspection procedures are also implemented. An important laboratory effort is aimed to the study of oxides, hydrides, corrosion and mechanical properties. Ferrero and Bavaro have developed numerical codes for simulating the channel in service distortion and fitting experimental measurements.

4. PREDICTING OF CHANNEL ELONGATION

For the sake of space, we shall limit ourselves here to describe the application of an old model of growth and creep developed by Savino and Laciana and later modified by Savino and Harriague for simulating fuel cladding deformation. The model simply applies a rate theory approach developed by Brailsford and Bullough for the point defects fate under irradiation. High texture is assumed and only two kinds of sinks for the defects: prismatic dislocations and grain boundaries. Recent work by A.M. Monti et al. for calculating the sink strengths in Zr is used; the main prediction of that work is that the strength depends on the D_c/D_a ratio for vacancies and interstitials. Defect dynamics computer simulation should predict interstitial preferential diffusion to prismatic dislocations with respect to basal ones and grain boundaries. The vacancy diffusivity is used as a variable parameter for each reactor type; that is, this parameter is fitted to one measured value while the others result from the model. Values of the different microstructural parameters used in the model are reported in Table I. We do not think that the model actually reproduces the physical reality but it should be taken as a framework that allows to include in a consistent numerical way the existing difference in metallurgical parameters between the materials used at either kind of reactor. This approach, if consistent with the experimental evidence, is therefore much better than a fully empirical one of measurements fitting.

Table I: Parameters used for the model

Parameters	CNA I	CNE
ρ	10^{12}	7×10^{14} m^{-2}
d_j	18×10^{-6} m	35×10^{-6} m (d_A)
(grain size)		0.7×10^{-6} m (d_R)
E_{fv}	1.8 eV	18×10-6 m
E_{mv}	1.47 eV	
D_{0v}	2.2×10^{-4} m^2/s	2×10^{-6} m^2/s
(fitted value)		
Texture	0.119 c//A	on R-T plane:
1.8 eV	0.881 c//A	0.6 at 80°C from R
		0.2 at 70°C from R
		0.2 at T

In Table II and Figure 1 the predictions of the model are compared with measured elongations at the CNE PT and CNA-I Coolant Channels; one of the measured elongations is taken as unity in each case. The values of prismatic dislocation density, grain shape and size and texture (discrete orientation of representative grains) used in each case are summarized in the Table I. Also, the vacancy diffusivity value obtained by fitting the measured channel elongation is reported.

Table II   Elongation Measurements (in units of L-09) elongation) - CNE- Zr 2.5 % Nb

Channel	Calculated Fast Neutron Flux (Fink)[1]	Measurement[2]	Ibrahim and Holt Prediction[2]	Own Work[2]
L-09	2.41	1	0.98	1.04
O-14	2.45	1	0.98	1.04
N-03	2.17	0.91	0.87	0.98
L-13	2.42	1.02	0.98	1.04
O-08	2.46	1	1	1.04
Q-09	2.41	1.02	0.98	1.04
E-16	2.23	0.91	0.89	0.98
N-07	2.46	1.02	0.98	1.04
F-09	2.33	1.02	0.94	1.02
O-02	1.69	0.74	0.68	0.83
W-10	0.95	0.45	0.38	0.59

[1] [10^{17} n/m^2 sec]   [2] $(\Delta l/l_0) / (\Delta l/l_0)$ L-09

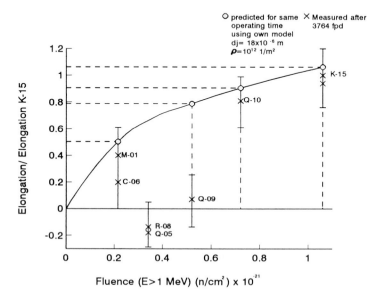

Fig. 1. Elongation Measurements CNA-I Zry-4 coolant channel

5. CONCLUSIONS

We have summarized in this work some of the efforts performed at CNEA to ensure the optimum performance of its nuclear plants. Fuel Channels at the CNE and Coolant Channels at the CNA-I and CNA-II are the vital inside parts of these reactors that constitute an eventual important life limiting factor. By a joint effort of operators and engineering and research groups, we aim to obtain the maximum efficiency and availability of the plants.

Experiments and theory provide the necessary basis for understanding Zr alloys performance at the reactor core. Modelling must be based on a sound basic understanding of the main microstructural variables that determine the performance. This understanding provides a unique way not only to a trustable extrapolation of the measurements at the reactor, but also to the planning of relevant experiments, which in turn provide tools for improving the component performance. The use of semiempirical models like the one discussed in Section 4 provides a way of consistently including the microstructural parameters of the material to fitting "in reactor" measurements and predicting the performance of reactor components.

ACKNOWLEDGEMENTS

The author wishes to thank Dr. A. Sarce for performing the calculations reported in Table II and Figure 1 and the utility operators (Gerencia de Area Centrales Nucleares, CNEA) for allowing the use of the results there
reported.

REFERENCES

Brailsford A D and Bullough R 1972 J. Nucl. Mat. 44 434
Ferrero A M J and Bavaro M A 1989 Private communication
Field G J 1988 J. of Nucl. Mat. 159 3
Fink J 1985 CNEA Internal Report 1083/85
Harriague S, Coroli G and Savino E J 1980 Nucl. Eng. and Design 56 91
Ibrahim G F and Holt R A 1980 J. Nucl. Mat. 91 311
Monti A M, Sarce A, Smetniansky-De Grande N, Savino E J and Tomé C N 1991 Phil. Mag. 63 925
Savino E J and Harriague S 1985 "Effects of Radiation in Materials" 12th Int. Symp. of ASTM, Philadelphia pp 667
Savino E J and Laciana C E 1980 J. of Nucl. Mat 90 90
Terlisky S 1988 Private communication
Tomé C N 1991 Private communication
Urbanic V F, Choubey R and Chow C K 1990 Proc. Zr in the Nuclear Industry- 9th Int. Symp., Kobe Japan
Urbanic V F, Cox B and Field G J 1987 Proc. Zr in the Nuclear Industry-7th Int. Symp., ASTM STP 939, Philadelphia pp 189
Urbanic V F, Warr B D, Manolescu A, Chow C K and Shanahan M W 1988 Proc. Zr in the Nuclear Industry- 8th Int. Symp., San Diego, California

Part 3
Modelling Fracture

Non-linear effects at crack tips

E Smith

Manchester University-UMIST Materials Science Centre, Grosvenor Street, Manchester M1 7HS.

ABSTRACT: Many brittle materials exhibit a behaviour, whereby the crack tip stress intensity increases during crack growth. This increase is associated with the formation of a ligament zone behind the crack tip, the restraining stresses due to the crack-bridging elements within this zone being responsible for the stress intensity increase. The paper focusses on the geometry dependence of the K-R curve, since this is important in the context of transferring crack growth data from cracked laboratory-type specimens to engineering component behaviour.

1. INTRODUCTION

My first contact with Ron Bullough was when we were Bruce Bilby's first research students at Sheffield. Ron was undertaking theoretical research on the structure of an edge dislocation, allowing for non-linearity of material behaviour in the tensile region beneath the dislocation. To celebrate his 60th birthday and to honour his contributions to materials science, it is appropriate that my paper should be concerned with non-linearity of material behaviour - however, the paper will discuss non-linearity in the vicinity of a crack tip rather than a dislocation.

The motivation for the research is the recognition that there are many types of brittle material, for example particulate reinforced ceramics, composites with brittle matrices, rubber-toughened brittle polymers and ceramics themselves, which can be toughened via a general mechanism which is based upon the restraining effects of material elements that bridge the faces of a crack. Upon loading a pre-cracked solid, the crack tip remains stationary until the crack tip stress intensity attains a critical value K_{IC} when the material fractures at the crack tip. With continued loading, the crack extends, leaving behind ligaments which bridge the crack faces, and the restraining stresses due to these ligaments are responsible for the stress intensity increasing during crack growth, and thereby introducing a desirable measure of toughness into the material. The opening at the original crack tip location, i.e. at the trailing edge of the ligament region, increases until the opening becomes too large for any bridging to occur, i.e. the restraint vanishes. In modelling this restraint for plane strain deformation, the simplest way is to smear-out the effect of individual ligaments and represent the restraint by an effective stress that restricts the relative displacement of the opposite faces of the crack.

The stress intensity (K) - crack extension (R) behaviour depends on a variety of factors: geometrical configuration, loading state, restraining stress-relative displacement law and the magnitude of K_{IC}; however the considerations can be simplified by assuming that $K_{IC} = 0$, and that the restraining stress retains a constant value p_c within the ligament region. The model is then similar to the DBCS model (Dugdale 1960, Bilby et al. 1963) of a plastically relaxed crack in a ductile material. Cottrell (1963) in his Tewksbury Lecture recognized the generality of this model of a crack tip with its associated constant stress zone at the crack tip; twenty five years later such a model can be seen to have a far wider range of applicability than was envisaged in the mid 1960's. For the idealised situation of a semi-infinite crack in a remotely loaded infinite solid, hereafter referred to as the "small zone" situation, Figure 1 shows a partially developed ligament zone of size R behind the crack

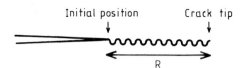

Figure 1 The model of a semi-infinite crack in a remotely loaded infinite solid.

tip, i.e. the crack has extended a distance R from its original position, the displacement at the trailing edge of this zone, i.e. at the initial crack tip, being δ. The stress intensity K at the crack tip due to the applied loadings, which is equal to the stress intensity due to the ligament material, is then given by the expression

$$K = p_c[8R/\pi]^{\frac{1}{2}} \tag{1}$$

this K-R relation being valid until the full development of the ligament zone, i.e. until δ attains a critical value δ_m. The critical K value associated with the attainment of a fully developed ligament zone is given by the expression

$$K_\infty = [E_o p_c \delta_m]^{\frac{1}{2}} \tag{2}$$

where $E_o = E/(1-\nu^2)$, E being Young's modulus and ν being Poisson's ratio. The fully developed ligament zone size R_∞ is given by the expression

$$R_\infty = \left[\frac{\pi E_o \delta_m}{8 p_c} \right]^{\frac{1}{2}} \tag{3}$$

Subsequent to the attainment of the critical fracture toughness K_∞, the crack continues to extend at the same applied stress intensity $K = K_\infty$ with a constant opening displacement being maintained at the trailing edge of the ligament zone, which retains a constant size. R_∞ is an important parameter because it provides an indication of the critical dimensions of a solid, e.g. initial crack size and initial ligament width, below which geometrical effects are expected to become important and the behaviour pattern appropriate to the small zone situation cannot then be used to describe the solid's behaviour.

For example, with a ceramic such as alumina where the restraining stress is due to the untangling of interlocking crystals, R_∞ is of the order of a mm (Majumdar et al. 1988), and geometrical effects are important with small laboratory specimens. On the other hand with a concrete-like material (Carpinteri 1989) or a cellulose/asbestos fibre reinforced mortar (Foote et al. 1986), R_∞ is of the order of tens of cms. and geometrical effects are then important even with large laboratory specimens. One theme of the author's research in this area of materials science during the last 2-3 years has been the effect of geometrical parameters on the K-R curve behaviour. This issue is important in the context of transferring crack growth data from laboratory-type specimens to engineering component behaviour, with the objective of taking credit for the enhanced toughness associated with ligament zone formation. The next section focusses on one particular example which highlights the importance of geometrical parameters.

2. THEORETICAL ANALYSIS FOR BEND SPECIMEN GEOMETRY

Most of the experimental crack growth resistance data for ligament toughened materials are obtained using bend specimens. We now, therefore, present the results of an analysis of the model (Figure 2) of a pure bend specimen of width w, thickness B, containing a crack with initial depth a_o, and initial ligament width b_o, the specimen being subjected to a bending moment M; there is a partially developed ligament zone of length R at the crack tip, which is therefore at a distance $a = a_o + R$ from the surface containing the crack.

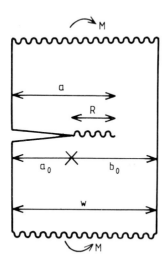

Figure 2 The model of a pure bend specimen subjected to a moment M.

The applied stress intensity, which is equal to the stress intensity due to the ligament material, is bounded (below) by the stress intensity for the situation where $a_o = 0$, and is bounded (above) by the stress intensity for the situation where $a_o = \infty$. These two extreme situations are easily analysed and the results are shown in Figure 3 along with those for the small zone situation (equation (1)); p_c is the uniform restraining stress within the ligament zone.

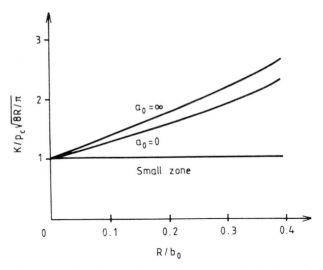

Figure 3 Bounding curves for the crack growth resistance behaviour of an edge-cracked bend specimen.

The conclusions that emerge from inspecting Figure 3 is that there is a marked difference between the resistance curve for an edge-cracked bend specimen and the small zone formulation.   There is a significant departure from the small-zone relation after only a

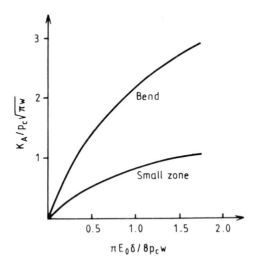

Figure 4 The relation between the bend specimen crack tip stress intensity K and the displacement δ at the initial crack tip.

small amount ($\sim 0.1\ b_o$) of crack growth, where b_o is the initial ligament width, although the dependence of the K-R curve on the initial crack depth is not especially pronounced.

Recognizing that the results in Figure 3 are with regard to the situation prior to the full development of a ligament zone, the results imply that the K-R curve is markedly geometry dependent, even with a material whose fully developed ligament zone size R_∞ according to the small zone formulation is small, i.e. with a ceramic where the restraining stress is provided by the untangling of interlocking crystals and R_∞ is of the order of a mm. The K value associated with a fully developed ligament zone in a finite width bend specimen increases as the initial ligament width decreases, and is markedly greater than the corresponding K value for the small zone situation. This is seen by referring to Figure 4, which shows the relation between the displacement δ at the initial crack tip and the crack tip stress intensity K for both the small zone situation and the edge-cracked bend specimen (for the case where the initial crack depth is vanishingly small). Proceeding from the basis that a ligament zone is fully developed when δ attains a critical value δ_m, it follows that the toughness associated with a fully developed ligament zone in a finite width bend specimen is markedly greater than the corresponding toughness for the small zone situation.

Thus it has been shown that the resistance curve for a ligament toughened brittle material is markedly dependent on geometrical parameters. In particular it has been shown that the general level of the K-R curve and the toughness level associated with a fully developed ligament zone is particularly high for the bend specimen. Caution must therefore be exercised in transferring bend specimen experimental data to say, for example, the behaviour of a surface edge crack in an engineering component, with the objective of taking credit for the toughness developed as a result of ligament zone formation; otherwise a non-conservative failure prediction may ensue. It is not the intention of this paper to indicate how this data transfer should be made, but rather to highlight the geometry dependence of the crack growth resistance curve, and indicate the problems that it can pose.

3. DISCUSSION

This paper has been concerned with one particular general problem area that involves the non-linearity of material behaviour at the tip of a crack, namely the modelling of the behaviour of a ligament toughened brittle material. The paper has addressed only one aspect of the behaviour of such a material: the geometry dependence of the crack growth resistance behaviour, and the problems this poses in attempting to take credit for the enhanced toughness that accrues as a result of crack extension. It should be pointed out that in this paper the restraining stresses provided by the individual ligaments behind a crack tip have been smeared-out, and an important question that must be asked is: "Under what conditions is this smearing-out procedure justified?" For the case of a semi-infinite crack in a remotely loaded infinite solid, and for the smeared-out situation, the crack extension condition can be expressed in the form

$$K = \left[E_o \int_0^{\delta_m} P(u)du \right]^{\frac{1}{2}} = \left[E_o \gamma_{EFF} \right]^{\frac{1}{2}} \tag{4}$$

where γ_{EFF} is an effective fracture energy and p(u) is the restraining stress when the relative displacement of the crack faces is u. It has recently been shown (Smith 1989), using a very simple model which allows for the discreteness of the individual ligaments, together with a general force-law describing the behaviour of an individual bridging element, that the smearing-out procedure can underestimate the toughening effect, i.e. the magnitude of γ_{EFF}, if the ligaments are brittle. There is an analogue here with the Peierls-Nabarro effect for a dislocation, where allowing for the discreteness of the crystal structure gives an enhanced value of the shear stress needed to move a dislocation. This is an appropriate point to draw this paper to its conclusion, for as indicated in the Introduction, the present author's first recollection of Ron Bullough was when he was undertaking his Ph.D. research on the non-linearity of material behaviour in the vicinity of a dislocation core.

REFERENCES

Dugdale, D.S., 1960, J. Mech. Phys. Solids, 8, p.100.

Bilby, B.A., Cottrell, A.H. and Swinden, K.H., 1963, Proc. Roy. Soc. A272, p.304.

Cottrell, A.H., 1963, Proceedings of First Tewksbury Conference, Faculty of Engineering, University of Melbourne, Australia, p. 1.

Majumdar, B.S., Rosenfield, A.R. and Duckworth, W.H., 1988, Eng. Fract. Mechs., 31, p.683.

Carpinteri, A., 1989, J. Mech. Phys. Solids, 37, p. 567.

Foote, R.M.L., Mai, Y.W. and Cotterell, B., 1986, J. Mech. Phys. Solids, 34, p.593.

Smith, E., 1989, Mechanics of Materials, 8, p.45.

Dynamic interface crack growth analysed by the method of caustics

K P Herrmann and A Noe

Laboratorium für Technische Mechanik, Paderborn University, Paderborn, F R G

Abstract: The equations of caustics for dynamically extending interface cracks are derived. An algorithm for the determination of stress intensity factors from experiments is obtained.

1. Introduction

In the scope of a fracture mechanical failure analysis of brittle composites the method of caustics is applied to determine stress intensity factors at crack tips situated in homogeneous components or at the interfaces of composites. The size of the caustics is affected by the mechanical and/or the thermal loading conditions, the curvature of the crack, the interface bonding and the rate of the crack tip velocity, respectively. In addition, the size of the caustics is influenced by the optical isotropy or anisotropy of the material of the associated material interface. The geometry of the caustics is proportional to the stress field gradient and therefore the caustic contour can be taken as a quantity for experimental measurements.
The method of caustics was developed by Manogg (1964) for a quasistatic crack propagation. Essentially, Theocaris (1976) extended the method to mixed-mode loading cases for interface cracks. The formulation of caustics equations for arbitrarily curved, quasistatic propagating interface cracks including optical isotropy and anisotropy, respectively, has been considered by Herrmann and Noe (1991) and Noe et al (1991). Experimental results and theoretically calculated caustics show that the application of the static theory instead of the dynamic theory in the case of higher crack tip velocities leads to essential differences. The first attempts to apply the method of caustics to fast running cracks have been performed by Kalthoff (1984) and Theocaris (1982). Extensions to mixed- mode loading cases of cracks in homogeneous materials including the consideration of the optical anisotropy have been carried out by Nishioka and Kittaka (1990).

2. Foundations of Caustics

Figure 1 shows the deviation of parallel light, penetrating a cracked two- phase specimen, due to the change of elasto- optical constants and the deformation of the specimen surface, induced by the stress field singularity which generates a shadow spot on the reference plane. The bright limit curve of the shadow spot is called caustic. The mapping equation for the shadow spot is given by the expression

$$\mathbf{w}_j = \mathbf{r}_j - C_j(\nabla(\sigma_1 + \sigma_2)_j \pm \lambda_j \nabla(\sigma_1 - \sigma_2)_j) \quad ; \quad (j = 1,2) \tag{1}$$

where $C_j = z_0 \, B \, c_j$ and c_j and λ_j are the shadow optical constants and the so- called optical anisotropy constants, respectively. The caustic can be obtained by introducing the so- called initial curve $r = r(\phi)$, following from the solutions of the eqs. (2), into eq. (1).

$$J = \frac{\partial(x', y')}{\partial(x, y)} = 0 \qquad \mathbf{w} = (x', y') \quad , \quad \mathbf{r} = (x \, (r, \phi), y \, (r, \phi)) \tag{2}$$

In the case of the existence of optical anisotropy in both parts of the composite two shadow spots and thus two caustics, so- called double caustics, are generated where the latter are build up from two initial curves. The caustics are no longer simply connected curves due to the discontinuity of the material constants along the interface.

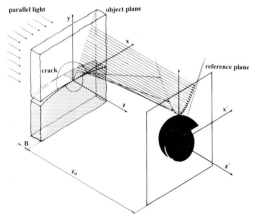

Fig. 1: Physical principle of caustics around an interface crack

3. Caustics for Dynamic Interface Crack Extension

In this chapter, the caustics equations for interface cracks running with a constant velocity v along the interface are derived where optical isotropy as well as optical anisotropy of the material is considered. The stresses may be related, according to Stroh (1962), to unknown generalized complex potentials, and can be derived by using a crack tip coordinate system running with velocity v. For linear elastic, isotropic material the following stress - potential relations are valid:

$$(\sigma_x + \sigma_y)_j = -(\beta_{1j}^2 - \beta_{2j}^2)\, Re\,(\Phi_{1j}(z_{1j})) \tag{3}$$

$$(\sigma_y - \sigma_x + 2\,i\,\tau_{xy})_j = 2\,(1+\beta_{1j}^2)\,Re\,(\Phi_{1j}(z_{1j})) + 2\,(1+\beta_{2j}^2)\,Re\,(\Phi_{2j}(z_{2j})) + \tag{4}$$

$$i\left(4\,\beta_{1j}\,Im\,(\Phi_{1j}(z_{1j})) + \frac{(1+\beta_{2j}^2)^2}{\beta_{2j}}\,Im\,(\Phi_{2j}(z_{2j}))\right) \quad ; (j=1,2)$$

$$\beta_{1j}^2 = 1 - \left(\frac{v}{v_{Lj}}\right)^2, \quad v_{Lj}^2 = \frac{\kappa_j+1}{\kappa_j-1}\frac{\mu_j}{\rho_j}, \quad \beta_{2j}^2 = 1 - \left(\frac{v}{v_{Tj}}\right)^2, \quad v_{Tj}^2 = \frac{\mu_j}{\rho_j} \tag{5}$$

where v_{Lj} and v_{Tj} are the longitudinal and the transversal wave velocities of material k (k=1,2), respectively, and z_{kj} is the generalized complex variable

$$z_{kj} = x + i\,\beta_{kj}\,y = r_{kj}\,\exp\,(i\,\phi_{kj}). \tag{6}$$

The desired potentials are obtained from the solution of the associated vectorial Hilbert- problem which arises from the boundary and continuity conditions of a stress- free interface crack. These potentials can be expressed as

$$\Phi_{kj}(z_{kj}) = \frac{1}{4\sqrt{2\,\pi}\,D_1 D_2 R_j}\,(F_{1kj}\,z_{kj}^{-1/2-i\beta} + F_{2kj}\,z_{kj}^{-1/2+i\beta}) \quad ; \quad (k,j=1,2) \tag{7}$$

where the constants D_1, D_2, R_j, F_{1kj}, F_{2kj} arise during the formulation and the solution procedure of the Hilbert- problem. Thereby F_{1kj} and F_{2kj} are linear functions of the stress intensity factors $K_{I,II}$. The constants D_1, D_2, R_j are related to the matrix

$$\mathbf{H} = \begin{pmatrix} -i\,D_1 & -w_3 \\ w_3 & -i\,D_2 \end{pmatrix} \quad , \quad \mathbf{H} = \begin{pmatrix} -i\left(\dfrac{B_{21}}{\mu_1 R_1} + \dfrac{B_{22}}{\mu_2 R_2}\right) & \left(\dfrac{S_1}{\mu_1 R_1} - \dfrac{S_2}{\mu_2 R_2}\right) \\ -\left(\dfrac{S_1}{\mu_1 R_1} - \dfrac{S_2}{\mu_2 R_2}\right) & -i\left(\dfrac{B_{11}}{\mu_1 R_1} + \dfrac{B_{12}}{\mu_2 R_2}\right) \end{pmatrix} \tag{8}$$

with

$$R_j = 4\,\beta_{1j}\beta_{2j} - (1 + \beta_{2j}^2)^2 \qquad\qquad B_{1j} = \beta_{1j}\,(1 - \beta_{2j}^2) \tag{9}$$

$$S_j = -2\,\beta_{1j}\beta_{2j} + (1 + \beta_{2j}^2) \qquad\qquad B_{2j} = \beta_{2j}\,(1 - \beta_{2j}^2)$$

where R_j is known as Rayleigh function. The non- trivial zero of R_j is called Rayleigh- wave velocity v_R. The Rayleigh- wave velocity v_R is ordered by $v_R < v_T < v_L$ where v_L and v_T are the minima of the longitudinal and transversal wave velocities in the eqs. (5), respectively.
The constants F_{1kj}, F_{2kj} can be calculated by using the matrices

$$\tilde{\mathbf{G}}_1 = \begin{pmatrix} \tilde{g}_{11}^{(1)} & i\,\tilde{g}_{12}^{(1)} \\ \tilde{g}_{21}^{(1)} & i\,\tilde{g}_{22}^{(1)} \end{pmatrix} \quad , \quad \tilde{\mathbf{G}}_1 = \begin{pmatrix} 4D_2\,\beta_{21} + 2w_3\,(1 + \beta_{21}^2) & i\,(4w_3\,\beta_{21} + 2D_1\,(1 + \beta_{21}^2)) \\ -4D_2\,\beta_{21} - 8w_3\,\dfrac{\beta_{11}\beta_{21}}{1 + \beta_{21}^2} & i\left(-4w_3\,\beta_{21} - 8D_1\,\dfrac{\beta_{11}\beta_{21}}{1 + \beta_{21}^2}\right) \end{pmatrix} \tag{10}$$

$$\tilde{\mathbf{G}}_2 = \begin{pmatrix} \tilde{g}_{11}^{(2)} & i\,\tilde{g}_{12}^{(2)} \\ \tilde{g}_{21}^{(2)} & i\,\tilde{g}_{22}^{(2)} \end{pmatrix} \quad , \quad \tilde{\mathbf{G}}_2 = \begin{pmatrix} 4D_2\,\beta_{22} - 2w_3\,(1 + \beta_{22}^2) & i\,(-4w_3\,\beta_{22} + 2D_1\,(1 + \beta_{22}^2)) \\ -4D_2\,\beta_{22} + 8w_3\,\dfrac{\beta_{12}\beta_{22}}{1 + \beta_{22}^2} & i\left(4w_3\,\beta_{22} - 8D_1\,\dfrac{\beta_{12}\beta_{22}}{1 + \beta_{22}^2}\right) \end{pmatrix} \tag{11}$$

and are given by the following expressions:

$$F_{11j} = (\tilde{g}_{11}^{(j)}\sqrt{D_1 D_2} - \tilde{g}_{12}^{(j)}D_2)\,K_I - i\,(\tilde{g}_{12}^{(j)}\sqrt{D_1 D_2} - \tilde{g}_{11}^{(j)}D_1)\,K_{II} \tag{12.a}$$

$$F_{21j} = -(\tilde{g}_{11}^{(j)}\sqrt{D_1 D_2} + \tilde{g}_{12}^{(j)}D_2)\,K_I + i\,(\tilde{g}_{12}^{(j)}\sqrt{D_1 D_2} + \tilde{g}_{11}^{(j)}D_1)\,K_{II} \tag{12.b}$$

$$F_{12j} = (\tilde{g}_{21}^{(j)}\sqrt{D_1 D_2} - \tilde{g}_{22}^{(j)}D_2)\,K_I - i\,(\tilde{g}_{22}^{(j)}\sqrt{D_1 D_2} - \tilde{g}_{21}^{(j)}D_1)\,K_{II} \tag{12.c}$$

$$F_{22j} = -(\tilde{g}_{21}^{(j)}\sqrt{D_1 D_2} + \tilde{g}_{22}^{(j)}D_2)\,K_I + i\,(\tilde{g}_{22}^{(j)}\sqrt{D_1 D_2} + \tilde{g}_{21}^{(j)}D_1)\,K_{II} \tag{12.d}$$

The stress intensity factors K_I and K_{II} are determined by the relation

$$\sigma_y + i\,\tau_{xy} = \frac{1}{2\sqrt{2\pi}\,\sqrt{D_1 D_2}}\,(\,((\sqrt{D_1 D_2} + D_2)\,K_I + i\,(\sqrt{D_1 D_2} + D_1)\,K_{II})\,r^{-1/2 - i\beta} +$$

$$((\sqrt{D_1 D_2} - D_2)\,K_I + i\,(\sqrt{D_1 D_2} - D_1)\,K_{II})\,r^{-1/2 + i\beta}) \quad . \tag{13}$$

The bimaterial constant β depends on the crack tip velocity v and reads

$$\beta = \beta(v) = \frac{1}{2\pi}\ln\left(\frac{\sqrt{D_1 D_2} + w_3}{\sqrt{D_1 D_2} - w_3}\right) \quad . \tag{14}$$

For the general case of optical anisotropy in eqs. (1) the stress gradients

$$\nabla\,(\sigma_1 + \sigma_2)_j = g_{1j} - i\,h_{1j} = -(\beta_{1j}^2 - \beta_{2j}^2)\,(\,Re(\,\Phi'_{1j}(z_{1j})) - i\,\beta_{1j}\,Im(\Phi'_{1j}(z_{1j}))) \quad ; \quad (j = 1,2) \tag{15}$$

and

$$\nabla \left(\sigma_1 - \sigma_2 \right)_j = g_{2j} - i \, h_{2j} = \frac{G_j}{N_j} - i \, \frac{H_j}{N_j} \tag{16}$$

$$G_j = f_{1j} f_{1j,x} + f_{2j} f_{2j,x} \quad , \quad H_j = -(f_{1j} f_{1j,y} + f_{2j} f_{2j,y}) \quad , \quad N_j = \sqrt{f_{1j}^2 + f_{2j}^2} \; . \tag{17}$$

$$\left| f_1 + i \, f_2 \right|_j = \left(\sigma_1 - \sigma_2 \right)_j = \left| \sigma_y - \sigma_x + 2 \, i \, \tau_{xy} \right|_j \tag{18}$$

have been calculated by using the stress potential- relations in eqs. (3) and (4). By introducing the eqs. (15) and (16) into eq. (1) the mapping function changes to the following notation

$$w_j = z_j - C_j (g - i \, h)_j \quad ; \quad (j = 1,2) \tag{19}$$

with

$$g_j = g_{1j} \pm \lambda_j \, g_{2j} \quad , \quad h_j = h_{1j} \pm \lambda_j \, h_{2j} \; . \tag{20}$$

For the sake of the determination of the initial curve the functional determinant (2) is calculated by using eq.(19) and reads by omitting the index j:

$$J = 1 + C \, u(x,y) - C^2 v(x,y) = 0 \tag{21}$$

where u and v are obtained from g and h as follows:

$$u = \frac{\partial h}{\partial y} - \frac{\partial g}{\partial x} \quad , \quad v = \frac{\partial g}{\partial x} \frac{\partial h}{\partial y} - \frac{\partial h}{\partial x} \frac{\partial g}{\partial y} \tag{22}$$

In the case of optical isotropy of the material the mapping equation and the functional determinant, respectively, are reduced to the following two expressions already given by Theocaris (1982).

$$w_j = z_j + 2 \, C_j \, (\beta_{1j}^2 - \beta_{2j}^2) \, (\, Re(\, \Phi'_{1j}(z_{1j})) - i \, \beta_{1j} \, Im(\Phi'_{1j}(z_{1j}))) \quad ; \quad (j = 1,2) \tag{23}$$

and

$$J = 1 + 2 \, C_j \, (1 - \beta_{1j}^2) \, (\beta_{1j}^2 - \beta_{2j}^2) \, Re(\, \Phi''_{1j}(z_{1j})) - (\, 2 \, C_j \, (\beta_{1j}^2 - \beta_{2j}^2) \, \beta_{1j} \, | \, \Phi''_{1j}(z_{1j}) \, |)^2 = 0 \tag{24}$$

The fundamental problem to calculate an initial curve $r = r(\phi)$ from eq. (24) can be only achieved in the case of a crack situated in a homogeneous material ($\beta = 0$) [6], otherwise the functional determinant (24) becomes a transcendental function and the initial curve has to be calculated numerically.

The caustics equations in the case of the optical anisotropy of the material ($\lambda \neq 0$) cannot be derived in an explicit manner, since lengthy expressions arise from which quantitative conclusions cannot be deduced.

Figures 2-4 demonstrate the influence of the crack tip velocity and the material inhomogeneity onto the shape of the caustics by consideration of different mode relations around the tip of a running interface crack where only the upper side of the material interface has been considered. The interface crack is assumed to be situated in a two-phase composite model with material properties of the combination Araldite B and steel. In addition, optical isotropy of the material ($\lambda = 0$) has been assumed. The discussion is based on the reference case of a quasistatically extending interface crack. The size of the caustics increases with increasing crack tip velocity, where this size enlargement is only significant for higher crack tip velocities compared to the minimum of the Rayleigh- wave velocities v_{R1} and v_{R2}. This result has been confirmed by the investigations of Kalthoff (1984) and Nishioka and Kittaka (1990) for running cracks in a homogeneous material. In view of the applicability of the method of caustics a strong sensitivity of the caustic geometry to mixed- mode conditions can be detected. Finally, in Fig. 5 both parts of a double caustic in the case of optical anisotropy of the material ($\lambda \neq 0$) are compared with the associated caustic for the case of optical isotropy. Fig. 5 shows that each branch of the double caustic is significantly different from the corresponding caustic in the case of optical isotropy. In addition, an overall view demonstrates that the influence of the optical anisotropy is more distinct for that chosen magnitude of λ.

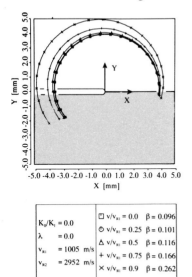

$K_{II}/K_I = 0.0$	$\square$ $v/v_{RI} = 0.0$ $\beta = 0.096$
$\lambda = 0.0$	$\circlearrowright$ $v/v_{RI} = 0.25$ $\beta = 0.101$
$v_{RI} = 1005$ m/s	$\triangle$ $v/v_{RI} = 0.5$ $\beta = 0.116$
$v_{R2} = 2952$ m/s	$+$ $v/v_{RI} = 0.75$ $\beta = 0.166$
	$\times$ $v/v_{RI} = 0.9$ $\beta = 0.262$

Fig. 2:   Mode-I caustics

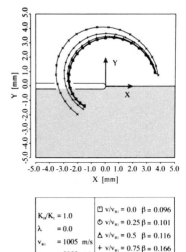

$K_{II}/K_I = 1.0$	$\square$ $v/v_{RI} = 0.0$ $\beta = 0.096$
$\lambda = 0.0$	$\circlearrowright$ $v/v_{RI} = 0.25$ $\beta = 0.101$
$v_{RI} = 1005$ m/s	$\triangle$ $v/v_{RI} = 0.5$ $\beta = 0.116$
$v_{R2} = 2952$ m/s	$+$ $v/v_{RI} = 0.75$ $\beta = 0.166$
	$\times$ $v/v_{RI} = 0.9$ $\beta = 0.262$

Fig.3:   Mixed-mode caustics

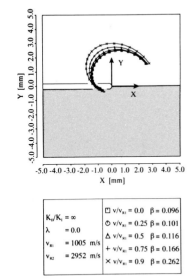

$K_{II}/K_I = \infty$	$\square$ $v/v_{RI} = 0.0$ $\beta = 0.096$
$\lambda = 0.0$	$\circlearrowright$ $v/v_{RI} = 0.25$ $\beta = 0.101$
$v_{RI} = 1005$ m/s	$\triangle$ $v/v_{RI} = 0.5$ $\beta = 0.116$
$v_{R2} = 2952$ m/s	$+$ $v/v_{RI} = 0.75$ $\beta = 0.166$
	$\times$ $v/v_{RI} = 0.9$ $\beta = 0.262$

Fig. 4:   Mode-II caustics

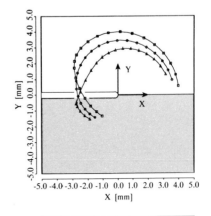

$\beta(v) = 0.116$	$\square$ $\lambda = 0.29$
$v/v_{RI} = 0.5$	$\circlearrowright$ $\lambda = 0.0$
$K_{II}/K_I = 1.0$	$\triangle$ $\lambda = -0.29$

Fig. 5:   Mixed-mode caustics

Figs. 2-4:   Influence of crack tip velocity and material inhomogeneity on the shape of caustics by consideration of optical isotropy of the material

Fig. 5:   Influence of optical anisotropy and material inhomogeneity on the shape of the caustics around a running interface crack tip

4. Experimental Determination of Stress Intensity Factors

Based on the caustics equations derived a measuring algorithm has been formulated in order to determine stress intensity factors $K_{I,II}$ from experimentally gained caustics. For a more accurate and more complete recording of the caustic geometry the application of a digital image processing technique has been included into the evaluation process, since the latter allows the usage of numerous appropriate measuring points along the caustic contour.

The caustics equations can only be formulated in an explicit manner if the functional determinant (2) can be solved for the initial curve $r = r(\phi)$. This fundamental requirement can only be achieved in the case of a crack in a homogeneous material ($\beta \neq 0$), otherwise the functional determinant becomes a transcendental function and the corresponding equation has to be solved numerically. It has to be pointed out that in addition to the stress intensity factors K_I and K_{II}, the radii r_k and the angles ϕ_k, respectively, of the measuring points in the reference plane are additional unknowns. The algorithm is using the two components of the mapping equation (1) and the functional determinant (2) where the latter has to be used necessarily since it selects the caustic as a special mapping function. For the algorithm a set of N ($N \geq 2$) measuring points $(x_k'^{(exp)}, y_k'^{(exp)})$ is needed in order to formulate a system of 3 N nonlinear equations for the set of (2 N + 2) unknowns K_I, K_{II} and r_k, ϕ_k (k = 1, N), respectively. It reads

$$\mathbf{f}^T = (\mathbf{f}_1, \cdots, \mathbf{f}_k, \cdots, \mathbf{f}_N) = \mathbf{0}^T \quad , \quad \mathbf{f}_k = \begin{pmatrix} x_k'^{(exp)} - (r_k \cos\phi_k - C\ g(K_I, K_{II}, r_k, \phi_k)) \\ y_k'^{(exp)} - (r_k \sin\phi_k + C\ h(K_I, K_{II}, r_k, \phi_k)) \\ 1 + C\ u(K_I, K_{II}, r_k, \phi_k) - C^2\ v(K_I, K_{II}, r_k, \phi_k) \end{pmatrix} . \quad (25)$$

By applying a series expansion this overdetermined system of nonlinear equations can be reduced to an overdetermined system of linear equations, which can be solved by a procedure for linear least-square problems.

The measuring algorithm has been tested for interface cracks by consideration of optical isotropy of the material. The stress intensity factors which have been calculated from a set of measuring points recorded from simulated caustics have been compared with the stress intensity factors assumed for the associated simulation. The proposed algorithm has provided results for K_I and K_{II} with an accuracy of at least 5 percent.

A more detailed outline concerning the algorithm and the results can be found in a forthcoming paper of Herrmann and Noe (1991).

5. REFERENCES

Herrmann K P and Noe A 1991 Submitted for publication in Engng. Fracture Mech

Kalthoff J F 1984 Optical Methods in Mechanics, 3. Static and Dynamic Photoelasticity and Caustics, Recent Applications (CISM: Udine)

Manogg P 1964 PhD Dissertation Ernst- Mach- Institut Freiburg, F R G

Nishioka T and Kittaka H 1990 Engng. Fracture Mech. 36 pp 987-98

Noe A, Ferber F and Herrmann K P 1991 In: Experimentelle Mechanik in Forschung und Praxis 14. GESA- Symposium VDI Berichte 882 pp 313-24

Stroh A N 1962 J. Math. Phys. 41 pp 77-103

Theocaris P S 1976 Acta Mechanica 24 pp 99-115

Theocaris P S 1982 Optical Engng. 21 pp 581-600

Acknowledgement - The financial support of the German Research Association (DFG) is gratefully acknowledged by the authors.

Interfacial cracks in composite materials

V.K. Tewary

Materials Reliability Division, National Institute of Standards and Technology, Boulder, CO 80303

Robb Thomson

National Institute of Standards and Technology, Gaithersburg, MD 20899

ABSTRACT: Atomic displacements are calculated in a bimaterial composite solid containing an interfacial crack by using the lattice statics Green's function method. A simple model of the composite is assumed which consists of two simple cubic lattices with nearest neighbor interactions and bonded along a cube surface. The atomic displacements, as calculated by using lattice statics, do not have the unphysical oscillations predicted by the continuum theory.

1. INTRODUCTION

The stress distribution in a bimaterial composite solid containing an interfacial crack has been calculated by Williams (1959), England (1965), Rice and Sih (1965), and Willis (1971) by using the continuum model. Subsequently, this problem has been addressed by many authors (see, for example Rice 1988 for an excellent review and other references). The continuum model gives an oscillatory behavior of the displacement field and stress. The theory predicts that the displacement field over the crack surfaces oscillate out of phase, such that the crack surfaces interpenetrate each other and the crack closes upon itself. This behavior is obviously unphysical.

Another difficulty in the continuum theory is that a stress intensity factor cannot be defined in the same sense as for a homogeneous solid (see, for example, Thomson 1986 for a review). The conventional definition of the stress intensity factor is the $r=0$ limit of $r^{1/2}\tau$ where r is the radial distance from the crack tip and τ is the stress tensor. This limit does not exist for an interfacial crack in a composite solid. A possible definition of the stress intensity factor for interfacial cracks in composite materials has been given by Rice (1988).

We suggest that the unphysical theoretical predictions mentioned above, arise from the basic inadequacy of the continuum theory to represent a solid close to a defect. It has been shown earlier (Tewary 1973) that the continuum model is the asymptotic limit of the discrete lattice model and is not valid near a defect. It is important, therefore, to use a discrete lattice model to represent the solid near the crack.

In this paper, we describe the lattice statics model of an interfacial crack in a bimaterial lattice (bicrystal). We find that the atomic displacements near the interfacial crack are not oscillatory. Our results are based upon a particularly simple model of the solid which consists of two simple cubic lattices with nearest neighbor interatomic interactions. The lattice constants of the two lattices are assumed to be the same but their interatomic force constants are taken to be different. This model is an extension of the 'atoms on rails' model, used by Hsieh and Thomson (1973) for a crack in a homogeneous solid.

2. LATTICE STATICS GREEN'S FUNCTION AND ATOMIC DISPLACEMENTS

We consider the Born von Karman model of a simple cubic lattice with nearest neighbor interatomic interaction as shown in Figure 1. We consider a planar crack formed by breaking the bonds between the atoms across the plane of the crack.

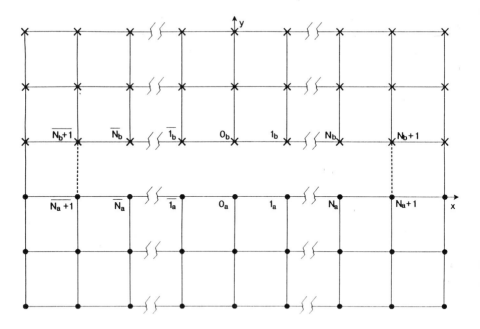

Figure 1 - Interfacial crack of length (2N+1) in the bicrystal AB consisting of two simple cubic lattices A and B. The crosses and dots denote, respectively, the lattice sites of solids B and A respectively. The subscripts b and a on atom numbers refer to solids B and A respectively. The absence of connecting lines show broken bonds. The dotted lines denote bonds which are broken first when the crack extends.

The basic parameter in lattice statics is the force constant matrix $\underline{\phi}$. The perfect lattice Green's function matrix $\underline{G}$ can be calculated in terms of the Fourier transform of $\underline{\phi}$ as described in Tewary (1973). The defect Green's function matrix $\underline{U}$ for a lattice containing a defect or a discontinuity is given by the solution of the Dyson's equation

$$\underline{U} = \underline{G} + \underline{G}\ \delta\phi\ \underline{U}, \tag{1}$$

where $\delta\phi$ denotes the change in the force constant matrix.

The main effort in lattice statics is in solving Eq. (1) for $\underline{U}$. This equation can be solved exactly if $\delta\phi$ is localized in space (Tewary 1973), i.e., if it is zero for all pairs of atoms except for a finite number. The vector space defined by the set of site vectors of atoms for which $\delta\phi$ is non-zero, is called the defect space. It is necessary to solve Eq. (1) only in the defect space. For a finite defect space, the solution of Eq. (1) is

$$\underline{U} = (\underline{I} - \underline{G}\ \delta\phi)^{-1}\ \underline{G}. \tag{2}$$

The matrix $\underline{u}$ of the atomic displacements induced by an applied external force $\underline{f}$ is given by (Tewary 1973)

$$\underline{u} = \underline{U}\ \underline{f}. \tag{3}$$

To solve the Dyson's equation, we need to localize $\delta\phi$ by using a suitable representation. We consider the final state of the solid for which the Green's function is to be calculated, as a defective solid in comparison to a suitable state taken as a reference state for which the Green's function is known. The defect is created by modifying the force constant matrix. Then, in Eq. (1), $\underline{G}$ is taken to be the known Green's function of the reference state, and $\delta\phi$ the change in the force constant matrix as measured from the reference state. The reference state is chosen judiciously such that $\delta\phi$ is localized. The Green's function $\underline{U}$ of the final state is then given by Eq. (2).

We consider a bicrystal AB made by two simple cubic solids, A and B. The interatomic interactions in the solid are restricted to first neighbors only. We construct the bicrystal AB in two steps. In the first step we cut the infinite solids A and B in two semi-infinite halves along the (0,1,0) plane. In the second step we join a half space solid A to a half space solid B along the cuts created in step 1. The first step creates free surfaces in A and B whereas the second step creates the required bicrystal. We calculate the Green's function for the two separated half space solids A and B by taking the perfect infinite solids as the reference state and the Green's function for the bicrystal by taking the two half space solids as the reference state.

Since the interface is along the XZ plane, the solid retains its periodicity in the X- and Z-directions. First, we reduce the problem to a 2D calculation by neglecting the variation in the Z- direction. Then we take the partial Fourier transform of each quantity in the

X- direction (Tewary 1973). These steps, combined with the nearest neighbor interaction assumption, localize the $\delta\phi$ matrix which enables us to solve the Dyson's equation. Thus we calculate the Green's functions for the two half space solids A and B in the first step and for the bicrystal AB in the second step. Our procedure for calculating the half space Green's function in the first step is similar to that used by Benedek (1978).

Finally, we create a crack at the interface by cutting the bonds between the atoms across the interfacial plane as shown in Figure 1. We represent the crack by equating the force constant matrix between atoms across the crack to zero (Hsieh and Thomson 1973). This defines the $\delta\phi$ matrix with respect to the bicrystal AB which is taken as the reference state. For a crack of finite length, the matrix $\delta\phi$ will be localized. We use this $\delta\phi$ in Eq. (1) and take $\underline{G}$ to be the Green's function for the bicrystal calculated in step 2 described above. Equation (2) then gives $\underline{U}$, the Green's function for the bicrystal containing the interfacial crack.

We now apply equal and opposite forces in the Y-direction on the two atoms at the center of the crack. The atomic displacements are calculated by using Eq. (3). The calculated atomic displacements of atoms over the crack surfaces for a crack of length $2N+1 = 21$ are shown in Figure 2. For these calculations we have assumed the following values of the force constants:

(i) bicrystal

$$\lambda_A = 2, \ \mu_A = 1, \ \lambda_B = 8, \ \mu_B = 3, \ \lambda_C = 5, \ \mu_C = 2, \tag{4}$$

(ii) homogeneous solid (average force constants)

$$\lambda_A = \lambda_B = \lambda_C = 5, \text{ and } \mu_A = \mu_B = \mu_C = 2, \tag{5}$$

where λ and μ denote, respectively, the radial and the tangential force constants between the first neighbors, and the subscripts A, B, and C denote, respectively, the lattices A, B and the bicrystal AB. Figure 2 also shows the atomic displacements for an average homogeneous solid with the force constants given by Eq. (5).

We see from Figure 2 that the atomic displacements in the homogeneous solid are symmetrical about the crack axis but not for the bicrystal. We also see that the atomic displacements in the bicrystal containing the interfacial crack vary smoothly and are not oscillatory as predicted by the continuum theory.

The displacements of the atoms over the crack surface in solid B are smaller in magnitude than the corresponding result for the homogeneous solid and vice versa for solid A. This is because the force constants for solid B and solid A are, respectively, larger and smaller than those for the homogeneous solid. A larger force constant implies a tougher spring and therefore its displacement for the same force is smaller. The crack opening is larger for the bicrystal than that for the homogeneous solid. The magnitude of the atomic displacements and the crack widening depends on the magnitude of the force constants.

Following Hsieh and Thomson (1973), we define a critical separation Δu_c between two atoms beyond which their interatomic bond will break. The crack opening force F_+ is defined as the force for which the separation between the atoms at the crack tip exceeds Δu and the crack opens by one atomic spacing. Similarly, the crack healing force F_- is defined to be the force at which the displacement between the two atoms at the crack tip falls below the critical value and the crack length decreases by a lattice length. The difference between F_+ and F_- is a measure of the lattice trapping of the crack (Hsieh and Thomson 1973).

Figure 3 shows the calculated values of $F_+/\Delta u_c$ and $F_-/\Delta u_c$ along with the corresponding values values for the homogeneous solid for which the force constants are given by Eq.(5). We see from Figure 3 that F_+ and F_- for the bicrystal are also proportional to $\sqrt{(2N+1)}$ as for a homogeneous solid (Hsieh and Thomson 1973).

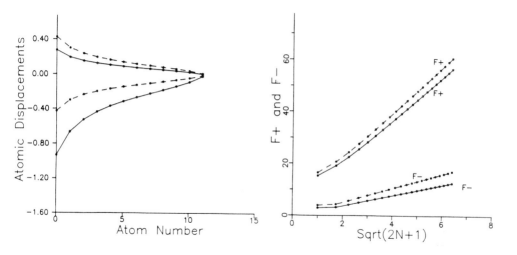

Figure 2 - Displacements of atoms on the positive X axis over the upper and the lower surface of the interfacial crack of length $2N+1 = 21$. The two end points denote the atoms adjacent to the crack tip which have unbroken bonds. Solid lines denote results for the bicrystal AB and dashed lines for a homogeneous solid for which the force constants are average of A and B.

Figure 3 - The crack opening (F_+) and healing (F_-) forces as functions of crack length. Solid lines denote results for the bicrystal AB and dashed lines for a homogeneous solid for which the force constants are average of A and B.

3. SUMMARY AND DISCUSSION

We have shown that, according to the lattice theory, the displacement field in a bicrystal containing an interfacial crack does not have the physically unrealistic oscillations which are
predicted by the continuum theory. Although the simple model which we have used is rather unrealistic, it does predict qualitative features of many properties of real solids (Tewary 1973, Hsieh and Thomson 1973). We believe, therefore, that our results are at least qualitatively correct.

The continuum theory is the asymptotic limit of the lattice theory (Tewary 1973) and is not applicable in a region close to a defect in a solid. It is not surprising, therefore, that the continuum theory predicts an unphysical result near the crack tip. Rice (1988) has suggested that the displacement and the stress field given by the continuum model are not valid close to the crack tip because of the non-linear effects but can be used to describe the state of the solid far away from the crack tip. This suggestion is consistent with our results on lattice statics. Our detailed lattice statics calculations on interfacial cracks will be published elsewhere.

REFERENCES

Benedek R 1978 J. Phys. F: Metal Phys. 8 1119
England A H 1965 J. Appl. Mech. 32 400
Hsieh C and Thomson R 1973 J. Appl. Phys. 44 1051
Rice J R and Sih GC 1965 J. Appl. Mech. 32 418
Rice J R 1988 Trans. of the ASME 55 98
Tewary V K 1973 Adv. in Phys. 22 757
Tewary V K, Wagoner R.H. and Hirth J.P. 1989 J. Mat. Res. 4 124
Thomson R 1986 Solid State Phys. 39 1
Williams M L 1959 Bull. Seismol. Soc. America 49 199
Willis JR 1971 J. Mech. Phys. Solids 19 353

High-temperature brittle intergranular fracture

H Rauh* and C A Hippsley

AEA Technology, Harwell Laboratory, Oxfordshire OX11 0RA, England
*Also Department of Materials, University of Oxford, Oxfordshire OX1 3PH, England

ABSTRACT: High-temperature brittle intergranular fracture is a novel type of crack growth which depends strongly on local sulphur segregation near crack-tips. It is most prominent in 'as-quenched' microstructures of alloy steels containing stress concentrators, loaded at temperatures above 300°C. Theoretical modelling of this fracture phenomenon has provided an understanding that enables corrective action to be taken against structural failure of service components in important fields of modern technology.

1. INTRODUCTION

Fracture studies constitute a cross-disciplinary technical area that involves both physicists and engineers. Physicists explore the basic processes giving rise to crack nucleation and development, while engineers quantify the resulting failures in actual plant components. Developing an understanding of the physics behind real engineering problems has the potential to generate ideas with many industrial applications.

One such application concerns the phenomenon of high-temperature brittle intergranular fracture (HTBIGF), a type of crack growth which causes intergranular failure of steels accompanied by very little micro-ductility. Discovered in the late 1970's, HTBIGF was originally identified as one of two mechanisms responsible for 'stress-relief cracking' during post-weld heat treatment of thick-section welds of ferritic steels, the other mechanism being the more conventional intergranular microvoid coalescence (Hippsley *et al.* 1982). Later, it was found to contribute to 'stress-relief cracking' in austenitic welds. Moreover, HTBIGF has received considerable attention as a failure mechanism in its own right, also governing slow, stable crack growth in unrelieved welds loaded at temperatures below the normal creep cracking regime (Smedley *et al.* 1984). This fracture mode, for instance, is thought to be primarily responsible for recent cracking in the superheater and reheater vessel shells of the Prototype Fast Reactor at Dounreay, which are made from 300 series austenitic stainless steel.

Initial laboratory evaluations of HTBIGF referred largely to ferritic steels. Both stress-relaxation and isothermal crack growth tests were performed on simulated and real heat-affected zone microstructures of low-alloy materials such as 2¼Cr1Mo and ½CrMoV or MnMoNi steels commonly used for pressure vessel fabrication (Hippsley *et al.* 1982, Hippsley 1983). Low-ductility intergranular crack growth, with fracture-surface morphology shown in Fig. 1(a), could be sustained in bend specimens subject to stress concentrators at temperatures between 300 and 650°C, and was found to be associated with carbide precipitation (providing potential crack nuclei) together with the segregation of phosphorus to grain-boundaries under thermal activation alone (Hippsley *et al.* 1984, Hippsley 1987). However, local enrichment of dynamically segregated sulphur in the region of high stress near crack-tips, depicted in Fig. 1(b), was identified as the most significant feature and the actual key to the operation of HTBIGF (Hippsley *et al.* 1982, Hippsley *et al.* 1984).

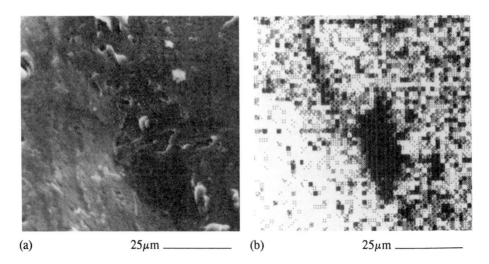

(a)   25μm _____   (b)   25μm _____

Fig. 1. Characteristics of HTBIGF in a commercial 2¼Cr1Mo steel. (a) Scanning electron microscope fractograph revealing the low-ductility nature of the intergranular fracture mode, (b) scanning Auger microprobe map indicating the distribution of sulphur (dark spots) close to the intergranular crack-tip.

Following this recognition, we proposed a fracture mechanism to explain the phenomenon of HTBIGF, based on the concept of solute segregation driven by the stress field of a loaded intergranular crack interacting with nearby point-defects. At temperatures high enough to ensure sufficient mobility, this interaction can impose a drift flow upon the point-defects that determines their migration in the vicinity of the tip. Thus, enhanced segregation to the crack as well as to the grain-boundary ahead of the crack may occur. When the defects are an embrittling solute (such as sulphur dissolved in an iron matrix during the high-temperature excursions imposed by welding or laboratory heat treatment) their accumulation can, in turn, promote fracture by reducing grain-boundary cohesion and encouraging crack propogation under the prevailing conditions of stress intensity and temperature (Hippsley *et al.* 1984, Hippsley 1987, Rauh *et al.* 1989).

In this paper we take the opportunity to highlight some of the recent progress achieved in modelling, verification and applications of HTBIGF. Correlations of model assumptions and theoretical predictions with experimental observations will be outlined and discussed in the light of available evidence. Attention will be drawn to the technological benefits obtained from our understanding of the new fracture mode, and comment made on further topical developments.

2. THEORETICAL MODELLING

Consider a long, straight, semi-infinite crack within an isotropic elastic body and a coplanar, unbroken grain-boundary ahead of the crack. An applied uniaxial tension produces mixed-mode loading on a scale local to the crack-tip due to the meandering process of intergranular fracture addressed, and the crack-tip exerts a stress field characterised by the stress-intensity factor K. When the axis of applied tension is orthogonal to the crack-tip and inclined at an angle α to the grain-boundary half plane (where $0 < \alpha < \pi/2$, without loss of generality), the energy of interaction between this stress field and a point-defect, represented by a misfitting spherical inclusion located at distance r from the tip and azimuth θ to the boundary, has the harmonic form (Rauh and

Bullough 1985, Rauh *et al.* 1989, Rauh and Bullough 1990)

$$E_\alpha = A \sin\alpha \, \sin(\theta/2-\alpha)/r^{1/2}; \quad -\pi < \theta < \pi, \quad (1)$$

in which

$$A = (2/9\pi)^{1/2}(1+v)K\Delta V, \quad (2)$$

where v denotes Poisson's ratio of the elastic body and ΔV the relaxation volume of the point-defect.

To analyse the kinetics of segregation, we envisage transient depletion of solute from an initial uniform solute atom concentration. Assuming that solute flow near the crack-tip is largely due to the crack-tip stresses and random diffusion may be ignored, the solute current density, in the 'pure-drift' approximation, obeys Einstein's equation (Rauh and Bullough 1985)

$$\mathbf{j}_\alpha = -(D/k_B T)c_\alpha \nabla E_\alpha \quad (3)$$

with the solute diffusion coefficient D, Boltzmann's constant k_B and absolute temperature T; the volume concentration of solute c_α itself is determined by the continuity equation

$$\delta c_\alpha/\delta t + \nabla \cdot \mathbf{j}_\alpha = 0 \quad (4)$$

recalling expression (3), together with the condition $c_\alpha = c_0$ adopted at initial time $t = 0$ and the requirement $c_\alpha = c_0$ as $r \rightarrow \infty$ at any time $t > 0$.

The appropriate solution for the transient solute atom concentration c_α around the crack-tip (implying that both the crack and the associated grain-boundary act as ideal point-defect sinks) consists of two regions separated by a characteristic which rigidly expands with time; in the inner region the concentration is identically zero and in the outer region it retains its initial value c_0. Fig. 2 illustrates this solution at a particular time $t > 0$ for $\alpha = \pi/3$ when $\Delta V < 0$, the case applying to sulphur solute in steels; the figure also shows equipotentials derived from eqn. (1) and flow lines given by orthogonal trajectories to the equipotentials along which solutes drift in the arrowed directions, thereby depleting the region inside the characteristic. Evidently, undersized solute atoms enter across the crack-surfaces and along the grain-boundary ahead of the crack, a prerequisite for embrittlement to take place within the fracture mechanism proposed. Furthermore, $N_\alpha(t)$, the total number of solute atoms lost at the various sinks after time $t > 0$, per unit length of the crack-tip in excess of the number deposited at time $t = 0$, is simply c_0 multiplied by the instantaneous area enclosed by the expanding characteristic (Rauh *et al.* 1989, Rauh 1990, Rauh and Bullough 1990).

To quantify solute segregation during intergranular crack propagation, we assume that crack growth proceeds in a 'step-wise' fashion, with average velocity v, by a succession of rapid jumps and stops: once embrittled locally through enrichment with sulphur at a level sufficient for decohesion of the unbroken grain-boundary adjacent to the tip, the crack jumps forward a discrete distance and then arrests in fresh (and hence tougher) material, waiting for further segregation during a time Δt, whereupon the process repeats. Provided that each jump takes the crack-tip well into regions where negligible depletion has occurred and the solute concentration is therefore at c_0, the average segregation can be deduced from the sequence of (small-time) depletions between jumps, when the crack is stationary. Using the fact that most of the grain-boundaries potentially suitable for intergranular crack propagation must needs lie reasonably close to the mode I loading orientation $\alpha = \pi/2$, the expected average total sulphur coverage on the fracture surface of an intergranular facet, per unit area of crack path in the mixed-mode loading

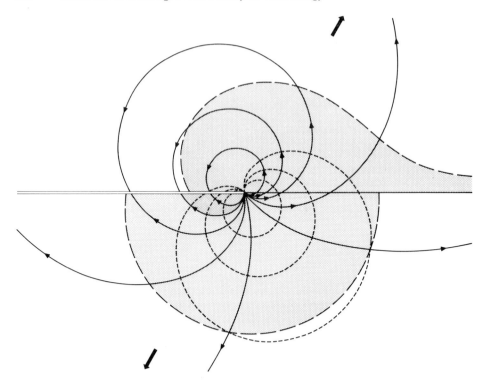

Fig. 2. Equipotentials (----), flow lines (——▶), and expanding characteristic (– –) in the vicinity of the crack (straight double lines) subject to mixed-mode loading by an applied uniaxial tension (bold arrows), and the unbroken grain-boundary (straight single line) ahead of the crack during transient depletion of solute atoms, when their relaxation volume $\Delta V < 0$. The solute concentration is zero inside the characteristic (shaded region) and equal to its initial value c_0 outside.

situation, is thus predicted to be

$$<P> \, = \, <N_\alpha(\Delta t)>/v\Delta t \sim N_{\pi/2}\,(\Delta t)/v\Delta t \tag{5}$$

with

$$N_{\pi/2}\,(\Delta t) = \frac{5}{2}\left[\frac{3\pi}{4}\right]^{1/5} c_0 \left[\frac{AD\Delta t}{k_B T}\right]^{4/5}, \tag{6}$$

i.e. essentially the same as if mode I loading and otherwise constant conditions prevailed (Hippsley *et al.* 1984, Rauh *et al.* 1989, Rauh 1990).

3. EXPERIMENTAL VERIFICATION

We now present a brief outline of the correlations of the various hypotheses underlying our theoretical model, and of the model's predictions, with experimental observations of HTBIGF in ferritic steels.

The implied assumption, that sulphur is the main embrittling agent triggering high-temperature crack growth, was validated in two ways, *viz.* through heat treatment and recourse to model alloys. First, a particular 2¼Cr1Mo steel was step-quenched from initial austenitisation at 1300°C. Successive quenches to 1200, 1100, 1000 and 900°C before final quenching to room temperature left progressively less sulphur in solid solution, while a constant average grain size was maintained. Isothermal crack growth tests revealed an increase of the propensity of this steel to HTBIGF with the quench temperature, and hence with the amount of dissolved sulphur available for segregation (Bowen and Hippsley 1988). Supplementary high-resolution scanning Auger microprobe analysis proved that sulphur taken into solution during austenitisation is the most important source of the segregating sulphur solute (Hippsley 1987). Secondly, a range of model alloys were prepared by doping the MnMoNi steel A533B with impurities of sulphur, phosphorus, boron and cerium, and isothermal crack growth tests were again performed. Fig. 3 shows the reduction of stress intensity required to maintain an average crack growth rate of $v = 3\mu\mathrm{ms}^{-1}$ at 550°C due to each dopant, or combination of dopants, when present at levels expected for commercial purity steels, measured relative to the stress intensity prevailing when doping is absent (Hippsley and Lane 1991). Clearly, sulphur, as a single dopant, is the most potent agent promoting HTBIGF in the aforementioned steel. It is interesting also to note that cerium, which is known to interact strongly with sulphur to form precipitates, and thus remove it from solid solution, has an extremely beneficial effect, counteracting the influence of both sulphur and phosphorus doping.

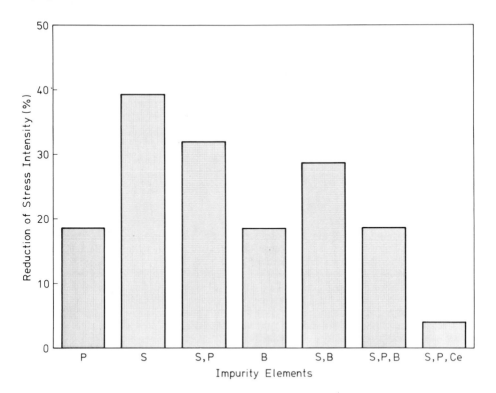

Fig. 3. Experimentally determined reduction of stress intensity required to sustain HTBIGF in the MnMoNi steel A533B as a result of doping with impurities. The doping level for sulphur and phosphorus is 300 ppm, for boron 4 ppm, and for cerium 350 ppm.

Likewise, the assumption, that intergranular crack propagation takes place in a 'step-wise' fashion, was validated by two types of technique, *viz.* fractography and acoustic emission measurements. Isothermal crack growth experiments were carried out on several ferritic steels under high-vacuum conditions in order to preserve the resultant fracture surfaces from loss of fine detail through oxidation (Bowen and Hippsley 1988). High-resolution scanning electron microscopy revealed striations spaced 1-10μm apart on fracture surfaces of intergranular facets (Fig. 4(a)); observed in profile, these striations manifested themselves as ridges of plastically deformed material (Fig. 4(b)). Both findings support the hypothesis that crack arrest and re-initiation had occurred after each local fracture event, and simultaneously quantify the range of jump distances involved. The fractographic evidence was subsequently confirmed by measurements of acoustic emission from samples undergoing HTBIGF (Hippsley *et al.* 1988). Discrete signals were obtained, again suggesting a 'step-wise' mode of crack growth. Signal amplitudes detected were approximately half of those measured for truly brittle intergranular fracture occurring at room temperature due to temper embrittling effects, and at least one order of magnitude greater than those derived from microvoid coalescence, demonstrating the essentially brittle nature of HTBIGF. The average crack jump distance derived from the rate of acoustic emission events observed relative to the overall crack growth rate was found to be in accord with the fractographic evidence.

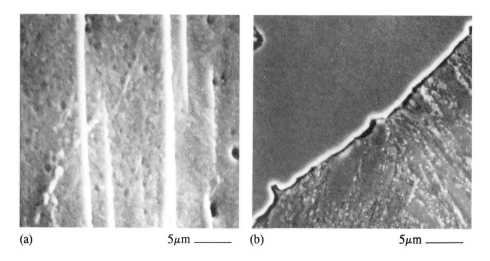

(a)   5μm _____   (b)   5μm _____

Fig. 4. Fractographic evidence of 'step-wise' crack growth during HTBIGF in a commercial 2¼Cr1Mo steel. (a) Fracture surface of an intergranular facet showing striations, (b) same fracture surface in section showing ridges.

The model predictions of total amounts of solute segregated to fracture surfaces during HTBIGF were verified by Auger electron spectroscopy. Notched bend specimens of 2¼Cr1Mo steel were austenitised at 1200°C and quenched leaving substitutionally dissolved sulphur atoms in solution. They were then fatigue pre-cracked and stressed, prior to Auger electron spectroscopical analysis of the resulting solute distribution near the growing HTBIGF crack-tips. Fig. 5 shows the sulphur coverage on fracture surfaces, eqn. (5), as a function of temperature predicted for the average crack propagation velocity $v = 0.1\mu ms^{-1}$ and the crack residence time $\Delta t = 1s$, using an empirical relation between crack growth rate and stress intensity, and appropriate values of the material parameters concerned (Hippsley *et al.* 1984, Rauh *et al.* 1989, Rauh 1990). We note excellent agreement with the measured data over the whole range of coverage, corresponding to 0.2 - 20% monolayer sulphur deposition on, for instance, a (100) crystallographic plane. Comparative free-surface segregation experiments lent further support to the assumption that this deposition arose predominantly from the action of the crack-tip stress field on the migrating sulphur solute during the relevant (short) time of crack arrests between crack jumps, rather than from diffusional effects (Hippsley *et al.* 1982, Hippsley *et al.* 1984, Hippsley 1987).

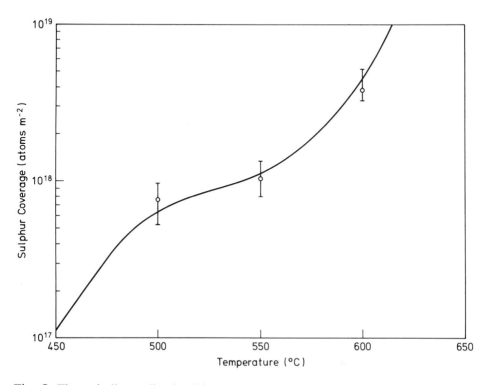

Fig. 5. Theoretically predicted sulphur coverage on fracture surfaces in a commercial 2¼Cr1Mo steel as a function of temperature (——), together with the available experimental data (o) and typical error bars.

Finite element calculations, extending the analysis of the preceding section to a blunt notch with an associated mode I elastic/plastic stress field interacting with undersized point-defects, have been performed on the basis of the 'pure-drift' approach. Resultant total point-defect flows were predicted very much like those in the purely elastic case, but with the origin of the flow lines centred at the position of (bounded) maximum hydrostatic stress displaced by approximately twice the notch-root radius in front of the physical notch-tip (Sinclair and Hippsley 1985, Rauh *et al.* 1989). Thus, sulphur atoms dissolved in a ferrite matrix would be driven away from the peak hydrostatic stress location towards the notch-root, and grain-boundary traps intersecting this high flux region enriched with sulphur and therefore embrittled. Studies of HTBIGF employing notched bend specimens of ferritic steels revealed that cracks do indeed nucleate in the said region, confirming the model prediction (Hippsley and Lewandowski 1988).

4. TECHNOLOGICAL APPLICATIONS

Following the identification of HTBIGF as a novel type of intergranular crack growth, the theoretical analysis described has provided an explanation of this phenomenon in ferritic steels, that is both qualitatively and quantitatively consistent with the experimental evidence to date. The model, presented here for undersized migrating point-defects with sulphur being the decisive embrittling solute, has been reformulated appropriately for the case of oversized point-defects and used to interpret hydrogen-induced environmental cracking of high-strength steels at room temperature, with similar success (Rauh 1990). Thus, it seems now well established that stress-driven solute segregation near crack-tips plays a key role for the mechanical behaviour of these materials, where performance depends on bulk transport of embrittling agents to prospective fracture zones. The understanding gained has paved the way for corrective actions taken on several occasions of failure due to HTBIGF in service components of nuclear reactor and conventional power plant, including the prevention of 'stress-relief cracking' in pressure vessel welds, and the modification of material and heat-treatment specification to minimize the occurrence of leaks in steam generator tube to tube-plate welds.

More recently, austenitic structural components operating at temperatures between 500 and 550°C have begun to manifest slow, low-ductility intergranular cracking in welds, with characteristics very similar to those of HTBIGF. Our knowledge of this fracture mode is still developing, but from studies of model alloys and service failures it currently appears that sulphur enrichment of the crack-tip region here too plays an important role. However, the phenomenon in austenitic steels differs from that in ferritic steels insofar as a plastic strain rate is also required (Ortner and Hippsley 1991). In addition, there is evidence to suggest that austenitic steels become more susceptible to HTBIGF after thermal ageing, with service failures encountered hitherto initiating after several years' exposure at elevated temperature. It is hoped that further systematic investigations will enable us to predict the high-temperature fracture behaviour of austenitic plant with the same confidence as is now possible for ferritic structures.

ACKNOWLEDGMENTS

Ron Bullough steadfastly encouraged the cross-disciplinary development of our understanding of the fracture mode described, and generously provided us with advice from his own depth of expertise. We take great pleasure in thanking him for his support. This work was carried out as part of the UKAEA Programme on Corporate Research.

REFERENCES

Bowen P and Hippsley C A 1988 *Acta Metall.* **36** 425
Hippsley C A 1983 *Metal. Sci.* **17** 277
Hippsley C A 1987 *Acta Metall.* **35** 2399
Hippsley C A, Buttle D J and Scruby C B 1988 *Acta Metall.* **36** 441
Hippsley C A, Knott J F and Edwards B C 1982 *Acta Metall.* **30** 641
Hippsley C A and Lane C E 1991 AEA Res. Rep. RS 2096
Hippsley C A and Lewandowski J J 1988 *Metall.Trans.* **19A** 3005
Hippsley C A, Rauh H and Bullough R 1984 *Acta Metall.* **32** 1381
Ortner S R and Hippsley C A 1991 AEA Res. Rep. TRS 4076
Rauh H 1990 in *Thermal Effects in Fracture of Multiphase Materials* eds K P Herrmann
    and Z S Olesiak (Berlin: Springer-Verlag) pp 46-56
Rauh H and Bullough R 1985 *Proc. R. Soc. London* **A397** 121
Rauh H and Bullough R 1990 *Proc. R. Soc. London* **A427** 1
Rauh H, Hippsley C A and Bullough R 1989 *Acta Metall.* **37** 269
Sinclair J E and Hippsley C A 1985 Presented at the *ORNL Second Int. Conf. on
    Fundamentals of Fracture*, Gatlinburg
Smedley J A, Broomfield A M and Anderson R 1984 *Nucl. Eng. Int.* **29** 26

Modelling the brittle–ductile transition

P.B. Hirsch and S.G. Roberts

Department of Materials, Oxford University, Parks Road, Oxford OX1 3PH, UK

ABSTRACT: A brief review is presented of a model for the brittle-ductile transition (BDT) of precracked single crystal specimens of materials which have a sharp transition (e.g. Si, Al_2O_3). At a given strain rate the transition temperature (T_c) is controlled by the speed with which crack tip sources can be nucleated by motion of existing dislocations in the bulk to the crack tip or at which loops emitted from special sites at the crack tip travel along it nucleating new sources along the whole of its length. At the BDT the nucleated sources send out avalanches of dislocations which shield the crack tip before the applied stress intensity factor K reaches K_{1c}. The competition between slip and cleavage at the crack tip is controlled by dislocation velocity relative to strain-rate.

1. INTRODUCTION

What are the mechanisms which control the brittle-ductile transition (BDT)? Rice and Thomson (1974) considered the mechanism of nucleation of a shear loop at the crack tip, and estimated activation energies for various materials, but an alternative mechanism is blunting and shielding due to the motion and multiplication of existing dislocations near the crack tip.

Even for the simplest materials there are different types of transition for precracked specimens:
1) Sharp transitions, in which the failure mode changes within a few degrees from brittle to ductile, with a large increase in failure stress (e.g. Si, Al_2O_3);
2) Gradual transitions in which the stress at which brittle fracture occurs increases gradually, perhaps over a range of 100°K or more, but without any stable crack growth (SCG) (e.g. Si after certain predeformation);
3) Gradual transition as in (2) but in which SCG occurs before brittle fracture (e.g. b.c.c. metals and alloys, MgO).
In this paper we shall review the dynamic shielding model which has been developed to explain the sharp transitions found in Si and ceramics.

2. THE BRITTLE-DUCTILE TRANSITION IN SILICON

Brede and Haasen (1988) and Michot and George (1986, 1989) have used precracked double cantilever type specimens of 'dislocation-free' silicon. The Oxford group (Samuels and Roberts 1989) have used four point bend type specimens, containing small surface cracks introduced by indentation which provides a source of external dislocations in the plastic zone near the surface. All the investigations, whatever the specimen geometry, agree on

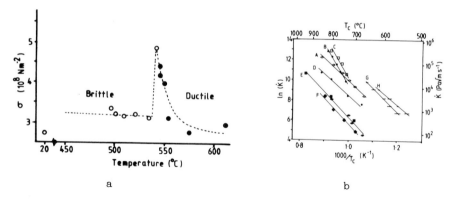

Figure 1a Failure stress against temperature for intrinsic silicon;
o brittle cleavage, ● ductile failure.
Figure 1b Brittle-ductile transition temperatures (T_c) at various stress
intensity factor rates ($\dot{K}$) in silicon. A-C, Brede and Haasen (1989); D,
St.John (1975); E-F, Michot and George (1986, 1989); G-H, Samuels and
Roberts (1989). For compositions see Hirsch and Roberts (1991).

the sharp nature of the transition (i.e. the transition range is only a few
degrees) (Figure 1a), and that the transition temperature T_c is strain-rate
dependent.   Below T_c the fracture is entirely brittle and the fracture
stress is independent of temperature; it is controlled through the Griffith
criterion by the low temperature critical stress intensity factor for
brittle fracture, K_{1C}.   Above T_c the whole specimen deforms plastically at
a stress determined by the upper yield stress.   At T_c there is a jump in
the stress intensity factor at failure (K_F) from K_{1C}, the low temperature
critical stress intensity factor, to a higher value, the maximum of which
is limited by the upper yield stress.   In this narrow transition region the
failure is essentially brittle, but occurs at values of $K_F \gg K_{1C}$.

Figure 1b shows the strain-rate sensitivity of T_c, expressed in the form of
$\ln(\dot{K})$ as a function of $(T_c)^{-1}$, where $\dot{K}$ is the rate of increase of the
stress intensity factor in a constant strain-rate test.   The two curves for
the Oxford experiments refer to intrinsic and n-type material respectively.
Writing

$$\exp(-U_{BDT}/kT_c) = C \dot{K} \tag{1}$$

where U_{BDT} is the activation energy controlling the strain-rate dependence
of T_c, k is the Boltzmann constant, and C is a constant, we find that U_{BDT}
is equal to the activation energy U controlling dislocation velocity (v):

$$v = A\tau^m\exp(-U/kT) = (\tau/\tau_o)^m v_o \tag{2}$$

where A and m are constants, τ is the applied stress and τ_o is a reference
stress.   Figure 1b shows clearly however that the intercepts of the various
curves, i.e. the values of C, are very different.   In the Oxford
experiments a change of crack radius from 13μm to 37μm led to an increase
of T_c by 50K (Samuels and Roberts 1989).   It appears therefore that C and
therefore T_c are structure sensitive.

Etch pit studies (Samuels and Roberts 1989) have shown that under dynamic
loading conditions at T_c dislocations emanate from a few sources along the

crack front, and that these sources begin to operate only when $K{\equiv}K_o{\gtrsim}0.9K_{1C}$. Under static loading conditions dislocation activity begins at $K{\sim}0.35K_{1C}$. A 'warm prestressing effect' has been observed (Booth, Cosgrave and Roberts 1990), and etch pit studies of such specimens show that the strain rate dependence of T_c is controlled by dislocation generation during loading.

3. THE DYNAMIC DISLOCATION SHIELDING MODEL

The model assumes that dislocation loops are generated by sources at particular points on the crack front. The dislocations parallel to the crack tip move away with a velocity given by Equation(2), where τ is the resultant of the crack tip field stress, the image stress and a line tension term, and the dislocation interactions. The parameters A,U,m in equation(2) are taken from George and Champier (1979). The loop is assumed to expand parallel and perpendicular to the crack tip at the same speed. The position of the dislocations are determined at any K, which increases at constant $\dot{K}$. Calculations have been performed for mode III and mode I. At any time the shielding K_D of the dislocations (e.g. Thomson 1986) can be calculated. The effective local stress intensity factors at the source (K_{es}) and at the mid point Z between sources (K_{ez}) have been derived, where it is assumed that only dislocations which have moved past Z will cause shielding there. New loops are emitted whenever the crack tip field stress at a critical distance x_c from the crack tip is sufficient to expand the loops against the line tension/image stress and repulsion from previously emitted dislocations. Figure 2a shows K_{es} and K_{ez} as a function of time. The effective K at the source K_{es} first falls and then rises as each dislocation is emitted and moves away from the crack tip. In the model fracture is assumed to occur when $K_{ez}=K_{1C}$, and the corresponding value of K_F is shown in Figure 2a. K_F is then calculated as a function of temperature, for given values of K_N (the value of K at which the first dislocation is emitted after nucleation), and the distance between the source and Z. The transition is predicted to occur in the observed temperature range, but to be gradual. Sharp transitions occur only if it is assumed that crack tip sources are nucleated at $K{\equiv}K_o$ very close to K_{1C} and that these sources then operate at $K=K_N{\ll}K_o$. Under these conditions an avalanche of dislocations is emitted from the sources at K_o, and provided the emitted dislocations can reach Z before K reaches K_{1C}, effective

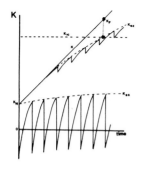

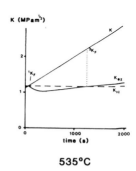

535°C

a   b

Figure 2a Variation with time of K, K_{es}, K_{ez}. Fracture occurs at $K=K_F$, when $K_{ez}=K_{1C}$.

Figure 2b Variation of K_{ez} at T_c. K_{ez} drops rapidly at $K={}^1K_F{\sim}K_{1C}$ due to emission of an avalanche of dislocations, and then rises gradually to reach K_{1C} when fracture occurs at ${}^2K_F$.

shielding occurs. Figure 2b shows the form of K_{ez} at T_c; K_{ez} deviates from K just before $K=K_{1C}$, then drops on account of the avalanche, and then increases gradually until $K_{ez}=K_{1C}$; the corresponding value of K is $K_F \gg K_{1C}$. A few degrees below T_c K reaches K_{1C} before K_{ez} deviates from K and the specimen fractures at $K=K_{1C}$. Thus there is a sharp jump in K_F at T, i.e. a very sharp transition is predicted in agreement with experiment (for examples see Hirsch, Roberts and Samuels 1989, Hirsch and Roberts 1991).

4. NUCLEATION MODELS

Model A. Effect of existing dislocations in the crystal

In the Oxford experiments the precursor cracks are introduced by surface indentations at room temperature. A plastic zone is formed in this process under the indentation. The nucleation event is considered to consist of the motion of dislocations from the plastic zone to the crack tip, forming crack tip sources there by a process of cross-slip. The details of the model are given in Hirsch et al (1989). The model predicts that for a given dislocation arrangement $\dot{K} \propto v_0$ at T_c as is observed experimentally (see Eqns(1,2)). The proportionality constant C in Eqn(1), is a function of the dislocation arrangement and is therefore structure sensitive. The agreement between theory and experiment is good. The model also accounts for the dependence of T_c on crack size, found in the Oxford experiments and referred to in Section 2. To check the assumption in the model that the pre-existing dislocations are in the indentation plastic zone at the surface, specimens have been tested in which the top $4\mu m$ were removed by polishing; T_c was found to increase by 50°C, in qualitative agreement with expectation. Abrading such polished surfaces decreased T_c again by 40°C. These experiments confirm that in the Oxford studies T_c is controlled by pre-existing dislocations at the surface (Warren 1989). The model also predicts that in specimens predeformed at T_c to K just below K_{1C}, crack tip sources should already be nucleated, and that such specimens should show a gradual transition extending to lower temperatures. This has also been confirmed experimentally (Warren 1989).

Model B. Low source density along crack front

In this model, which is believed to apply to the precracked double canti-lever type specimens of 'dislocation-free' Si (Brede and Haasen 1988, Michot and George 1986, 1989), dislocations are assumed to be emitted from a few special sites along the crack, such as at the surface or at ledges, at $K=K_N \ll K_{1C}$. Loops emitted from these sites help to activate other more difficult sites along the crack front because of the antishielding stresses ahead of these loops, or generate new sources by cross-slip of the advanc-ing screws. Each new source is assumed to operate at $K_N \ll K_{1C}$. Thus a mini-Lüders band propagates along and away from the crack front (Figure 3). Calculations have been carried out for different values of the original and final source separations, $2_i d_{crit}$, $2_f d_{crit}$ respectively. Shielding at point Z is assumed to occur only when loops have moved past Z. The condition for the BDT, and T_c, is that the loops from the original sources have travelled $_i d_{crit} -_f d_{crit}$ when K reaches K_0 (just below K_{1C}), such that the loops from the last nucleated sources have time to traverse $_f d_{crit}$ before K reaches K_{1C}, as shown in Figure 3(iii); $_f d_{crit}$ determines $K_{1C}-K_0$ and the sharpness of the transition. As in Model A an avalanche of dislocations is generated by the last source, and if $_f d_{crit}$ is small a significant drop in K_{ez} occurs and the transition is sharp (see Figure 2b). Detailed computations are shown in Hirsch and Roberts (1991). The model also predicts that at T_c, if $(_i d_{crit} -_f d_{crit})$ does not vary, $(\dot{K}/v_0)$ is constant, in agreement with experiment (see Section 2).

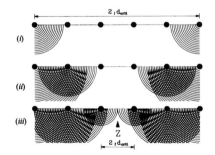

Figure 3 Nucleation model in which a few sources $2_i d_{crit}$ apart send out dislocations which in turn activate other sources ($2_f d_{crit}$ apart) when they reach them along the crack tip. Shielding at Z occurs when dislocations $_f d_{crit}$ from Z move past Z.

Table 1. Mode I simulations for Nancy experiments (Michot 1988, Michot and George 1989)

$\dot{K}/\dot{K}_s$	0.3	1.0	2.0	10.0	
T_c (exp)	946	988	1000	1075	(K)
τ (exp)	10	14.6	16.3	13.1	(MPa)
T_c (model)	945	988	1015	1090	(K)
τ (model)	10 - 11	10 - 12	10 - 12	9 - 14	(MPa)
v_d (model)	0.8	2.8	6	30	($10^{-7}ms^{-1}$)
$v_d/\dot{K}/\dot{K}_s$	2.7	2.8	3.0	3.0	($10^{-7}ms^{-1}$)

$$(\dot{K}_s = 310 \text{ Pa}\sqrt{m}s^{-1})$$

for the model $K_N = K_o = 0.3K_{1c}$, $K_{1c} = 0.93$ MPa$\sqrt{m}$, $_i d_{crit} = 550\mu m$

Michot and George (1989) have determined T_c at different $\dot{K}$, and have measured the velocities (v_d) of the outer edges of the plastic zone by X-ray topography. Table 1 shows their experimental results for various strain-rates $\dot{K}$ (relative to a standard strain rate $\dot{K}_s$); τ(exp) is the effective stress acting on the leading dislocations in the plastic zone deduced from the observed velocities v_d, and using Equation(2). We have carried out calculations for the known values of K_N, K_{1c}, and adjusting $_i d_{crit}$ to give the best agreement between observed and calculated values of T_c. The results are compared in the table; the agreement is surprisingly good, in view of the crudity of the model compared to the more complex geometry in the experiments. The value of $_i d_{crit}$ is of the order of half the specimen thickness, suggesting surface nucleation (Hirsch and Roberts 1991). More recently Azzouzi, Michot and George (1991) have confirmed from etch pit patterns of specimens which failed in a brittle manner at T_c corresponding to strain rates in the range of Table 1 that slip initiates at the surface. These workers also found that at very slow strain-rates a number of sources are activated along the crack front, i.e. $_i d_{crit}$ is smaller, and correspondingly T_c is lower than expected from the lines in Figure 1b. Computations on the model for such cases correctly predict T_c for values of $_i d_{crit}$ of the same order of magnitude as those observed. The model has also been used to calculate the velocity of the expanding plastic zone, as a function of K. The experimental values (Michot and George 1989) fall within the range of rates calculated using the mode I and mode III models (Hirsch and Roberts 1991).

5. DISCUSSION

The two-stage model of nucleation of crack tip sources followed by rapid shielding of the crack tip has proved successful in explaining the characteristic features of the sharp BDT in Si, and the gradual transition observed when the nucleation stage has been eliminated by suitable predeformation. In particular the model explains the dependence of T_c on strain-rate and on the initial distribution of dislocation sources in the bulk and/or on the crack tip, making the BDT a structure sensitive property. The model provides a description for the first two types of transition identified in Section 1, in which no stable crack growth occurs before brittle fracture. The model has also been applied successfully to sapphire, which also has a sharp transition similar to Si (Roberts, Kim and Hirsch 1991). It should be noted that the critical parameter which controls the competition between cleavage and slip at a crack tip is <u>dislocation velocity</u>. Even if dislocations are emitted from the crack tip at $K < K_{1c}$, as K increases the back stress from the emitted dislocations can become sufficiently great for cleavage to occur after some slip when $K - K_D = K_{1c}$.

The model does not account for stable crack growth. For this to occur dislocation sources must exist or be generated in the bulk ahead of the crack tip so that an extended plastic zone forms. Within such a zone stable crack growth can be promoted by various mechanisms - micro-cracking promoted by antishielding dislocations within the zone, cavity nucleation and growth around particles, punching out of dislocations from a blunted crack to annul antishielding dislocations, etc. With computer power now available it should be possible to model the plastic zone by slip bands passing through and ahead of the crack tip and, as K increases, to follow their growth, the stresses within the zone and the effective value of K at the crack tip. With such a model it may become possible to determine the amount of crack growth for a given value of K. In the meantime, Ellis (1991) has recently shown that in precracked single crystals of Mo in which stable crack growth is small compared to the length of the slip band, the gradual BDT observed can be modelled successfully.

REFERENCES

Azzouzi H, Michot G and George A 1991 *Proc. 9th Int. Conf. on Strength of Metals and Alloys* ed D G Brandon, R Chaim and A Rosen (London: Freund Publ. House) pp 783-789
Booth A S, Cosgrave M and Roberts S G 1990 *Acta Metall. Mater.* **39** 191
Brede M and Haasen P 1988 *Acta Metall.* **36** 2003
Ellis M 1991 D.Phil. Thesis, University of Oxford
Hirsch P B and Roberts S G 1991 *Phil. Mag.* A**64** 55
Hirsch P B, Roberts S G and Samuels J 1989 *Proc. Roy. Soc.* A**421** 25
Michot G and George A 1986 *Scripta Metall.* **20** 1495
Michot G and George A 1989 *Structure and Properties of Dislocations in Semiconductors* eds S G Roberts, D B Holt and P R Wilshaw (Bristol: Institute of Physics Press) IOP Conference Series 104 pp285-407
Rice J and Thomson R 1974 *Phil. Mag.* **29** 73
Roberts S G, Kim H S and Hirsch P B 1991 *Proc. 9th Int. Conf. on Strength of Metals and Alloys* ed D G Brandon, R Chaim and A Rosen (London: Freund Publ. House) pp 317-324
Samuels J and Roberts S G 1989 *Proc. Roy. Soc.* A**421** 1
St. John C 1975 *Phil. Mag.* **32** 1193
Thomson R 1986 *Solid State Physics* eds H Ehrenreich and D Turnbull (New York: Academic Press) **39** pp 1-129
Warren P D 1989 *Scripta Metall.* **23** 637

Strategies for structural integrity

B Tomkins

AEA Technology, Risley, Warrington, Cheshire, WA3 6AT

ABSTRACT: The assurance of integrity of engineering structures and components is important for both safety and economic considerations. Three approaches or strategies are employed; safe life, fail safe and damage tolerance. These strategies require a combination of operating experience and understanding of material and structural failure. Materials modelling is the basis of the utilisation of this understanding, particularly in damage tolerance. This strategy is used increasingly to complement the other more experience based strategies.

1. INTRODUCTION

All major industries today face the twin pressures of safety and economics, with regard to their operating plant. As far as the plant hardware is concerned, these pressures demand an assurance of structural integrity throughout life. The traditional source of such assurance is operating experience of similar plant, linked to appropriate technical knowledge and understanding. Together these inputs have been embodied in design and fabrication codes and standards, which evolve with an industry as it develops. Such codes and standards provide the basic level of integrity assurance, being augmented by operating rules and in-service examination as appropriate.

Three approaches on 'strategies' for structural integrity can be discerned across a wide range of engineering components and structures; safe life, fail safe and damage tolerance. The balance of experience and technical understanding varies in each but there is a clear historical trend to increase the input of technical understanding. This is in response to both stricter safety demands and the desire to maximise return on financial investment, for example by increasing a plant's performance or lifetime. The requirement is to quantify the level of risk which in turn requires a more quantified description of material and structural failure. Modelling provides the route for transferring scientific knowledge to application in engineering codes and rules, via an integrity strategy.

2. SAFE LIFE

Figure 1 shows the simplest framework for the integrity assessment of a component at the design stage. It is used in a wide range of design codes and standards from aeroengines to bridges to pressure vessels. It represents the SAFE LIFE approach. A, F and D represent three states of the component. A is the design state; that which is achieved if the rules for design and manufacture/fabrication are followed strictly. It does not claim to represent a state of perfection, but the rules attempt to ensure that initial imperfections ('original sin') will not seriously impair the intended strength or life of the component in operation. Definition of the failure state, F, takes account of the component type. It may be a local material failure at a stress concentration or weld over a volume of a few cubic mm or total structural failure of the whole component. Either way it represents the first unacceptable condition likely to be encountered by either overloading the component or running it to failure. It is the limit state. Once A and F have been defined, the design allowable state D can be defined such that operation within AD alone will ensure no failure - a SAFE LIFE. F/D represents the design margin. Typical examples of design margins are F/D = 20 for fatigue failure of a pressure vessel or 1.5 against excessive deformation. Where F can be defined by a statistical distribution, D is given by a low failure probability state, eg at -4 standard deviations for gas turbine engine components.

FRAMEWORK FOR THE INTEGRITY ASSESSMENT OF A COMPONENT

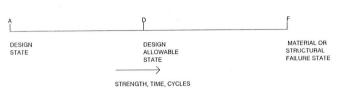

Figure 1

The safe life condition D is usually determined empirically and represents an acceptable condition in terms of operating experience. Similarly, the condition A is judged acceptable by practical experience in relation to the quality of design and fabrication consistent with the code. In terms of life, the position of D takes into account deficiencies in the design state A. However, uncertainties in A and F for an actual operating component make judgement on the adequacy of the acceptable design state D open to question if a high level of integrity assurance is required. This limits the SAFE LIFE strategy as sole guarantor of adequate integrity. Development of the safe life framework is needed in order to consider these limitations further and this is shown in Figure 2. Two new states (B, E) have been introduced as well as a more rigorous definition of F.

FRAMEWORK FOR THE INTEGRITY ASSESSMENT OF A COMPONENT
- INCLUDING THE 'BATHTUB' CURVE

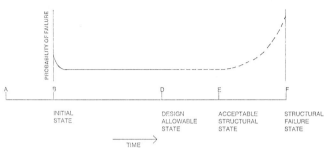

Figure 2

B represents the actual initial state of the component with respect to imperfections or defects, which could reduce the life, eg porosity, cracks. The scale of such flaws depends on material processing and component fabrication and can vary from a fraction of a mm to several cm in large structures. Considerable effort has been made to minimise initial imperfections by strict quality control and the integrity and associated reliability of many engineering components has increased considerably in recent years. The increased reliability of air and ground transportation vehicles bears this out. Increased reliability can be represented as a reduced failure probability in the early stages of a component's life. This is shown in Figure 2 as the first part of the well known 'bathtub' curve describing component reliability. As B moves towards D the initial failure probability increases, whilst the obverse is true as it approaches A. For critical components such as pressure vessels, an overpressure (overloading) test is carried out prior to service to eliminate vessels whose initial state is likely to be unacceptable.

The design allowable state D should always define an acceptable state of the component with regard to the development of failure. This means that the risk of failure at D is acceptable in terms of both plant safety and reliability. E represents the actual limiting acceptable state for a specific component. B and E are component specific states corresponding to the idealised states A and D defined by the design code. D must always lie to the left of E but if the margin DE is too large, a design will be inefficient. Beyond E, the development of material and structural damage increases the probability of failure beyond an acceptable level, EF representing an increasing risk of unacceptable structural failure.

The above description of failure development has been expressed in terms of component life, where the development of fracture damage (holes, cracks) may occur from pre-existing defects or occur naturally by the operation of time dependent failure processes such as fatigue, creep or stress corrosion. The SAFE LIFE strategy, based on an arbitrarily defined margin to failure (F/D) supported by experience, cannot provide a quantitative estimate of the likelihood of failure or remaining life of a given component. For components where such an estimate is required, a new strategy has evolved - DAMAGE TOLERANCE. This seeks, by non-destructive examination and quantitative knowledge of the rate of damage development and failure condition, to establish the actual state of a component at any point in its life. Before considering this strategy in detail, however, it is appropriate to consider a more simple engineered strategy for structural integrity, which is inherent to many components and structures - FAIL SAFE.

3.   FAIL SAFE

Most civil engineering constructions are highly redundant. Failure in one part of the structure can be tolerated because stresses are naturally redistributed in adjacent members. The existence of these alternative load paths ensures that total structural failure is gradual and is usually associated with excessive deformation. Detection of such deformation is a telltale sign that structural failure is occurring and steps can be taken to alleviate the problem. Most of us rely on this FAIL SAFE situation to repair our homes at appropriate times, thus prolonging their life as long as we wish. Where civil structures are not tolerant of deformation, for example modern thin walled dams which rely for their strength on maintenance of a given shape, catastrophic failure can be induced by significant ground movements. Steel frames structures, such as offshore oil platforms, are good examples of highly redundant structures which are tolerant of considerable local failure damage in individual members.

The essence of the FAIL SAFE strategy is to engineer naturally or by design a component or structure where local failure is tolerated and detected prior to catastrophic failure. In terms of Figure 2, it is not possible to reach F without the guaranteed detection of impending failure. The FAIL SAFE strategy has been widely used in the airframe industry by the provision by design of independent load paths which can be automatically brought into action if the primary path fails. Redundancy is a common concept in safety related control systems where multiple back-ups are introduced to ensure that system failure can be tolerated.

Lack of inherent redundancy makes the FAIL SAFE strategy inapplicable to pressure boundaries. Pressure vessels and pipework present a single section boundary between the vessel contents and the ambient environment. The integrity of the boundary is particularly important where the contents are toxic or dangerous if ejected. However, boundary failure can often be tolerated if the result is a small controlled leak. Most vessel failures do, in practice, involve a leak rather than a catastrophic rupture, or break. Statistical evidence is that the ratio of leaks to breaks for steel vessels is of order $10\text{-}10^2$. Leak-before-break is a FAIL SAFE concept and can be used as a structural integrity strategy for pressure boundaries if it can be rigorously demonstrated. It is most likely for vessels whose material remains ductile and whose failure is localised. This requires an understanding of the way in which failure develops in such components and the provision of adequate leak detection capability. As with SAFE LIFE, advanced FAIL SAFE strategies require quantitative knowledge to support them and cannot be simply engineered in.

4.   DAMAGE TOLERANCE

Both the SAFE LIFE and FAIL SAFE concepts rely on relating the state of a component at any time in its life to a known end state. The degree to which the known end state is relevant varies and uncertainties are swept up in the design margin (F/D). This is an integral approach where the integrated damage state of a component ('original sin' plus developed damage) is judged against an integrated damage state - failure. The considerable uncertainties in this approach for safety critical components and structures has been noted earlier and the third strategy, DAMAGE TOLERANCE, seeks to provide more certainty by adopting a differential approach, open to the powerful tools of material and structural modelling. The concern is to know the past or present state accurately, by means of non-destructive examination, and the limiting rate of damage development by all failure processes likely to be operating.

At the heart of the strategy is the understanding of material failure processes and an ability to relate them to a component. This understanding has increased strongly over the past 30 years with the advent of electron microscopes and advanced materials testing. Fracture mechanics, in its broadest sense, has provided the analytical modelling tool to quantify this understanding and make it applicable to components. Figure 3, after Ashby (1) shows the broad classes of fracture mechanisms in materials. Some lead to brittle and some to ductile behaviour and of particular structural interest are materials such as ferritic steels which can undergo a ductile/brittle transition, thus making design difficult for many components including pressure vessels.

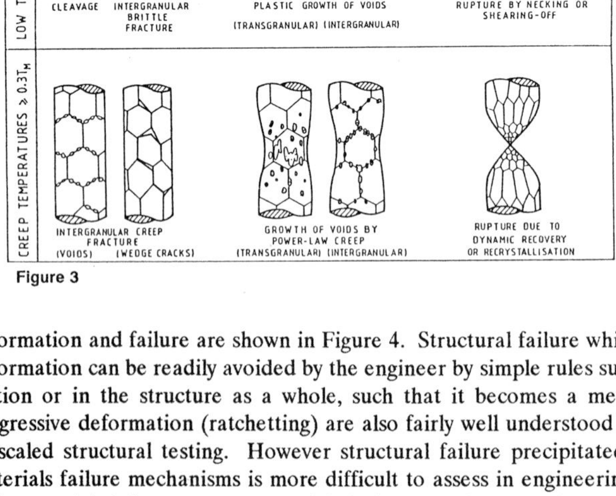

BROAD CLASSES OF FRACTURE MECHANISM

BRITTLE ←——————————————————————→ DUCTILE

LOW TEMPERATURES < 0.3T_M

CLEAVAGE | INTERGRANULAR BRITTLE FRACTURE | PLASTIC GROWTH OF VOIDS (TRANSGRANULAR) (INTERGRANULAR) | RUPTURE BY NECKING OR SHEARING-OFF

CREEP TEMPERATURES ≥ 0.3T_M

INTERGRANULAR CREEP FRACTURE (VOIDS) (WEDGE CRACKS) | GROWTH OF VOIDS BY POWER-LAW CREEP (TRANSGRANULAR) (INTERGRANULAR) | RUPTURE DUE TO DYNAMIC RECOVERY OR RECRYSTALLISATION

Figure 3

The links between material and structural deformation and failure are shown in Figure 4. Structural failure which results from excessive material deformation can be readily avoided by the engineer by simple rules such as avoiding gross yield within a section or in the structure as a whole, such that it becomes a mechanism. Buckling, collapse and progressive deformation (ratchetting) are also fairly well understood and designs can be cross-checked by scaled structural testing. However structural failure precipitated by the gradual development of materials failure mechanisms is more difficult to assess in engineering terms. Figure 3 shows that the basic materials failure processes result in holes or cracks on a micro- and progressively macro-scale. The time dependent failure processes of fatigue, creep and stress corrosion all result in crack or hole formation and holes, which are a result of the basic ductile tensile failure process in materials, link to form cracks prior to structural failure. The modelling of these processes to define their rate of development and limiting condition in terms of applied stress, strain, time, cycles is the key to quantifying structural failure.

To date, most modelling advance has been made in relating macroscopic failure parameters to microscopic features. For example, for ductile fracture, Ashby (1) was able to relate ductility (e_f), as a result of hole growth from inclusions, to the inclusion volume fracture (f_v), where,

$$e_f \approx \frac{1}{C} \ln \left\{ \alpha \left[\frac{1}{f_v^{1/2}} \right] - 1 \right\} \tag{1}$$

Where C and α are constants. Crack growth can be expressed in terms of a parameter (K, J) which characterises the crack trip stress-strain field where material failure is occurring. Empirical relations of the form,

$$\frac{da}{dt}^1 \frac{da}{dN}^1 \Delta a = f(J^*)_1 \, f(\Delta K)_1 \, f(J) \tag{2}$$

where J^* is a time dependent variation of J, can be derived for materials in such a form that they can be directly related to structures. Work is still needed to link microstructural modelling more clearly to these characterising parameters for structures. Meanwhile, structural modelling in the presence of cracks enables the limiting state of the characterising parameter (J_c, K_c), determined from materials tests, to be related to the structural failure condition, F.

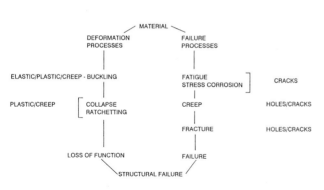

Figure 4

LINKS BETWEEN MATERIAL AND STRUCTURAL DEFORMATION AND FAILURE

Work remains to be done to cope with structural factors such as complex geometry and multiaxial loading, but fracture mechanics modelling has developed to a sufficient extent to provide an underpinning of quantified understanding to structural integrity assessment.

In addition to knowledge of the kinetics of failure development and the structural failure state, the DAMAGE TOLERANCE strategy requires a third input; knowledge of the state of the component or structure, prior to and probably during service. Diagnostics by means of non-destructive examination has been a major development of the last 20 years. It can range from visual to full volumetric ultrasonic examination. Again modelling plays a key role in translating the examination images into detailed descriptions of damage. Cracks must be not just identified but sized if their continued development is to be predicted accurately. More work is needed to strengthen the linkage of the key elements of the damage tolerance strategy in order to maximise its effectiveness at minimum cost.

5.   FUTURE DEVELOPMENTS

The Study of the integrity of the Pressurised Water Reactor pressure vessel carried out by a committee under the chairmanship of Lord Marshall (2) represented a milestone in integrity assessments for critical safety related components. The vessel was designed on a SAFE LIFE basis and a FAIL SAFE leak-before-break argument could not be made. Uncertainties in the initial state of this component made a DAMAGE TOLERANCE assessment necessary and this was made. A strict in-service inspection schedule linked to a fracture mechanics assessment provided the required added confidence to the SAFE LIFE design. The Study also outlined a probabalistic damage tolerance approach, based on the uncertainties associated with the inputs of non-destructive examination, stress analysis and materials response. Such probabalistic assessments are becoming more accepted as a complement to the usual deterministic DAMAGE TOLERANCE route. This enables risk to be predicted at both start of life and during life, thus enabling the 'bathtub' curve of Figure 2 to be evaluated.

INTERDISCIPLINARY REQUIREMENTS FOR STRUCTURAL INTEGRITY

ISSUE	ACTIVITY	DISCIPLINES	RESEARCH
1. STATE OF PLANT	- DIAGNOSTICS (INSPECTION, MONITORING, SURVEILLANCE)	● PHYSICS ● MATERIALS ● ENGINEERING	● MICROMEASUREMENT OF PROPERTIES ● DATA PROCESSING ● HUMAN FACTORS ● AUTOMATION ● ROBOTICS
2. ASSESSMENT/ REPAIR	- PREDICTION OF CONDITIONS AND KINETICS FOR FAILURE AND RECOMMENDED ACTION	● MATERIALS ● MECHANICS	● MATERIAL FAILURE MECHANISMS ● ADVANCED ANALYSIS ● 'SMART' EXPERIMENTS ● 'DESIGNER' REPAIRS

Figure 5

Plant which has been in operation for many years and is approaching its original design life is now having key components assessed by the DAMAGE TOLERANCE strategy to justify life extension beyond the design allowable state, D. The principle is as for new components: obtain a diagnostic fingerprint, relevant material and structural data and perform a deterministic and/or probabalistic assessment. In order to cope with the increasing use of DAMAGE TOLERANCE, models which underpin its use will need to be strengthened and validated. This requires an interdisciplinary approach with engineers and scientists working closely together. Figure 5 outlines the basic activities along with required disciplines and interdisciplinary research. 'Materials Modelling - From Theory to Technology' is a good description of the task which must be continued to assure structural integrity in the future.

6.   REFERENCES

1.   M F Ashby, Micromechanisms of Fracture in Static and Cyclic Failure, in Fracture Mechanics - Current Status, Future Prospects. ed R A Smith, Pergamon Press (1979).

2.   An Assessment of the Integrity of PWR Pressure Vessels, Report of a Study Group chaired by W Marshall, UKAEA - 1st Report 1976, 2nd Report 1982.

Part 4
Sensors and Electronic Behaviour

Modelling in automobile emissions abatement research

A D Brailsford

Research Staff, Ford Motor Company, Dearborn, Michigan, USA 48121-2053

Abstract: The feedback emissions control system in use in modern automobiles is described. Modelling in many different forms has played a vital role in its development. Some specific examples are highlighted.

1. INTRODUCTION

The past two decades have witnessed a dramatic but largely unheralded evolution in automotive technology. Spurred on by Federal legislation, motor vehicle manufacturers and their suppliers have developed (and mass-produced) emissions abatement systems that are able to reduce the three major pollutants in automobile exhausts by 96% in the case of total hydrocarbons and carbon monoxide or 76% in the case of oxides of nitrogen (designated as NO_x). A historical summary of progress is given in Table 1. In this brief review, we will present a simplified account of the means, in addition to improvements in engine technology, by which this success was achieved, with specific reference to the contributions of modelling to the development of the individual components of the emissions abatement system.

All production vehicles employ a basic feedback control system that was developed initially by Rivard (1973). It is comprised of four main elements: (1) an electronic fuel injection (EFI) unit for precise control of the relative mass of air to fuel (A/F) which is fed into the engine, (2) an exhaust gas oxygen (EGO) sensor to probe the exhaust gas A/F and detect stoichiometry in the incompletely combusted exhaust air-fuel mixture, (3) a three-way catalytic converter that simultaneously converts the reducing hydrocarbons (HC) and CO to H_2O and CO_2 (rich mixtures), or the oxidizing gas NO to N_2 (lean mixtures), and (4), a degree of exhaust gas recirculation (EGR) to curtail the formation of oxides of nitrogen (NO_x). Additionally, in order to produce a fully automated feedback control system, the on-board microprocessor (or electronic engine controller) is used to convert the sensed engine exhaust gas sensor and other signals, such as engine speed and load, into modifications of the engine inputs (A/F, ignition timing and EGR flow rate) such that the desired engine operating point is obtained.

Specifics of the components of the above will be discussed shortly. However, it can be appreciated already that there has been, and continues to be, a vital part for modelling to play in the development and continued refinement of this system. For example, vehicle and dynamometer tests, from which emissions data are obtained, are both expensive and time consuming. Thus, models which can to some extent predict the behavior of components (and, ultimately, the whole system) in both steady state and transient conditions, and in both new and aged systems (e.g. after of the order of a hundred thousand miles road service) yield significant economic as well as scientific benefit. Further, such modelling is seen to come in distinct forms, one of which concerns the mathematical description of the systems end of the spectrum. Another focusses upon the materials science properties of components such as the

Model Year	Hydrocarbons		Carbon Monoxide		Nitrogen Oxides	
	Grams	Reduction	Grams	Reduction	Grams	Reduction
Precontrol	10.6	-	84.0	-	4.1	-
1968-1971	4.1	62%	34.0	60%	NR	-
1972-1974	3.0	72%	28.0	67%	3.1	24%
1975-1976	1.5	86%	15.0	82%	3.1	24%
1977-1979	1.5	86%	15.0	82%	2.0	51%
1980	0.41	96%	7.0	92%	2.0	51%
1981-1982	0.41	96%	3.4	96%	1.0	76%
1983-1992	0.41	96%	3.4	96%	1.0	76%

SOURCE: U.S. Environmental Protection Agency and MVMA (1991) *Motor Vehicle Facts & Figures '91*
(Detroit: Motor Vehicle Manufacturers Association of the United States)

Table 1. Passenger Car/Exhaust Emissions Reduction Progress
Federal 49 State Standards (Grams Per Mile)

catalyst and the exhaust gas sensor. Clearly each activity has an indispensable part to play in the mathematical simulation of the behavior of the total system, both in its various modes of operation and at various stages of its service life.

2. THE CATALYTIC CONVERTER

Spark-ignited engines emit pollutant gases at concentrations varying with the intake air-fuel ratio as indicated in Figure 1. As a result, according to Kummer (1980), under urban driving conditions, cars without deliberate exhaust gas treatment, when calibrated for maximum fuel economy, give emissions in excess of the 1982 legislated levels. Some form of exhaust gas treatment is therefore necessary. The method of choice is the use of a catalytic converter (now referred to more simply as a catalyst). Catalytic oxidation of HC and CO in the exhaust can be   achieved for gas temperatures greater than ca. 300°C (the so-called

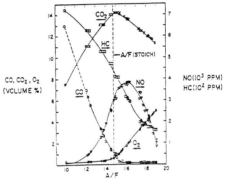

Fig. 1. Variation of Exhaust Emissions with Air/Fuel Ratio for Single Cylinder Engine using Indolene 30 Fuel (#1)

"light-off" temperature) by the use of Pt or Pd as a catalyst. The temperature limitation, apart from cold-start conditions, is perfectly acceptable. The removal of NO_x is achieved by reduction using the CO and HC present simultaneously in the exhaust. Rhodium is a good catalyst material for this purpose. The emission of NO_x is also suppressed by EGR. This process depletes the concentration of oxygen in the engine cylinder, thereby reducing the magnitude of the combustion heat generated, the peak cylinder temperature, and hence the concentration of oxides of nitrogen in the exhaust gas.

The single most important discovery in automotive catalysis has proved to be that, if an engine can be operated at or close to the stoichiometric point, it is possible to achieve conversion of all three pollutants using a single catalyst structure. Such a catalyst is called a three-way catalyst (TWC) because of this ability to remove the three emissions at one time. However, the three-way action obtains only over a very narrow range of A/F centered around stoichiometry. Thus, the challenge becomes one of how to sense that composition and how to maintain engine operation near it. These tasks define the roles of the exhaust gas oxygen sensor, and the electronic feedback control system, respectively, both of which are discussed in later sections.

The physical structure of the catalyst is that of a honeycombed brick (or monolith) with axial channels ~ 1 mm separated by walls ~ 0.1 mm thick. The walls themselves are also porous. The number of channels is around 400 per square inch and a number of different channel (or passageway) cross-sectional shapes are available. The actual catalytic material (usually noble metal particles on gamma alumina) is deposited on the surfaces of the channels in what is called a washcoat.

Modelling enters automotive catalysis in two distinct ways. One involves chemical or chemical engineering reaction kinetics and has the aim of determining conversion efficiencies for CO and selected HC's, given different conditions of temperature, gas flow rate and inlet concentrations of the exhaust gas constituents. The other is concerned with the materials science side of the subject. It addresses changes in physical microstructure; for example, the question of how surface noble metal particles might coarsen during service (analogous to bulk Ostwald ripening). Thereby, it relates to changes in individual chemical reaction rates, whereas the former is directed at the catalyst performance that these changes in physical structure (or particle dispersion) engender.

The chemical engineering modelling has by now been developed to a very sophisticated level. The latest version of the Ford TWC model, as described by Montreuil et. al. (1991), treats 13 independent forward paths for the oxidation of various exhaust gases with O_2 and NO as oxidizing species. The kinetic schemes contain 95 parameters (although only ~ 20 prove crucial): these are determined by numerous laboratory calibration experiments prior to using the model as a tool for exploring the effects of different catalyst formulations, for example. At the present time, extension of the model to transient conditions is under active development.

Examples of the materials science modelling in this technology are provided by the early work of Wynblatt and colleagues (1975) on supported particle coarsening via different modes of matter transport, and by a later study of Williamson et al. (1980) of the effect of surface segregation on the catalytic activity of Pt-Rh alloys. The issue in the first of these examples is the change in total noble metal particle surface area (or catalytic efficiency), i.e. aging, as a result of the coarsening of the distribution. The reduction in surface free energy is the driving force, just as in normal coarsening of second phase particles in bulk materials (see, for example, Brailsford and Wynblatt (1979)). In the second example, while the lowering of surface free energy is again the driver, the atomic mechanisms here involve atomic interchange of different constituents of an alloy and relative re-apportionment between surface and interior of the different alloy constituents. Further, in addition to the effects of like and unlike pair bonding in the solid state, the reactivity of each alloy with oxygen provides an additional driving force in an oxidizing environment. It should be emphasized that such noble metal alloys are not an intentional component of the catalyst, but may arise under high temperature reducing conditions through surface migration and chemical interaction of the individual noble metal constituents. In the optimum situation, one would like to have each precious metal supported on a specific inert (e.g. refractory oxide) support so as to maximize the catalytic efficiency per metal atom. Moreover, this situation has to be maintained over the life of the catalyst. It should also be noted that the support for each precious metal may not be the same because of strong metal - support interaction. This is a further complex area where significant modelling has reinforced experimental investigations.

Just as particle coarsening models can be used to assess aging, others can be formulated to describe catalyst poisoning (Oh and Cavendish (1983)). However, as Kummer points out, assessments so derived represent upper bounds to durability issues since vehicle malfunctions, for example, may be expected always to harm more than help. The random system malfunction, of course, lies outside the realm of the present discussion, though it is clearly of importance in modelling the real-life expectancy of components.

3. THE EXHAUST GAS OXYGEN SENSOR

It has been pointed out earlier that the three-way catalytic action is operative in only a small range (or "window") near stoichiometry in the exhaust gas mixture. It is imperative for efficient emissions abatement to be able to locate, or sense, that particular composition point. Such is the function of the Exhaust Gas Oxygen (EGO) sensor.

The predominant EGO sensor in use is an electrochemical cell made from yttria-stabilized zirconia. It is thimble-shaped and has porous platinum electrodes on its inner and outer surfaces. The inner surface is exposed to a standard atmosphere (ambient in the present instance) while the outer is inserted in the exhaust gas stream just beyond the exhaust manifold. In order to both protect the outer electrode and increase the residence time of the exhaust gas in this region, a porous spinel coats the outer surface. A metallic shroud with louvers or other orifices to allow gas circulation completes the structure. It is roughly the size of a conventional spark plug.

Many of the details of EGO sensor and its applications to the automobile emissions problem have been reviewed already by Logothetis (1981). Here we simply highlight the key points. The first of these is the means by which the stoichiometric A/F is detected: this is done via measurement of the partial pressure of oxygen in the exhaust. The change in oxygen partial pressure at stoichiometry is actually quite small unless the gases can be reacted locally to thermal equilibrium. This is the situation realized as a result of the following factors: (1) the catalytic action of the platinum electrodes (2) the enhanced residence time caused by the spinel and (3) in many cases, by the incorporation of an internal heater within the sensor thimble, which increases the gaseous reaction rates on the outer surface of the cell.

The atomic mechanisms by which the oxygen partial pressure is sensed involves transfer of oxygen between the host zirconia and the gaseous environment. Removal of a doubly charged oxygen ion, for example, is believed to occur at sites close to the platinum-zirconia-gas triple line, the two electrons from the ion being trapped in the platinum electrode, while the neutral oxygen atom migrates over the electrode surface, combines with another like entity, or an adsorbed carbon monoxide molecule, and then desorbs. Details have been presented by Fleming (1980) and Mizusaki et al (1987). This overall process (or its reverse) creates (destroys) a charged vacancy within the electrolyte. Similar processes occur at the reference electrode, the only difference between electrodes being the different adsorbed oxygen concentrations arising from the different surrounding oxygen partial pressures. The net result is a re-distribution of electronic and ionic (vacancy) charge across the sensor which produces a steady state emf given by the Nernst equation:

$$emf = (RT/4F) \ln [P_1 / P_2]$$

once the vacancy flux is zero. Here R and F are the gas and Faraday's constant and T is the absolute temperature: the P_2 are oxygen partial pressures in the two regions adjacent to the electrodes.

The results of laboratory tests of a commercial sensor placed in an engine exhaust are shown in Figure 2. The almost ideal step-like change of the output signal is, of course, a direct reflection of the variation in the oxygen partial pressure around stoichiometry. This is the prototypical behavior upon which control systems are based, since the sensor shows a clean sharp marker at stoichiometry, the location of the TWC window.

Unfortunately, there are many factors which tend to spoil this match. Most originate from the fact that the EGO is an open system in steady state rather than a closed system in thermal equilibrium. Brailsford and Logothetis (1985) have modelled several consequences of this distinction for sensor systems of this general type. With a representative set of interacting gases they showed that the output step could be virtually at any A/F value, only high temperatures and strongly catalyzed reactions locate it at stoichiometry. Earlier, Saji et al (1981) had showed that the spinel coating and/or the hydrodynamic boundary layer resulting from gas flow can also cause a shift in the step. So, too, can deposits on the sensor resulting from aging in the hostile exhaust gas environment. It is apparent therefore that the complete understanding, characterization and control of sensor properties represents a continuing challenge to experimental scientist, modeller and manufacturer alike.

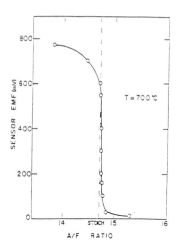

Fig. 2. Dependence of the emf of a ZrO_2 Oxygen Sensor on A/F at 700°C

4. THE FEEDBACK CONTROL SYSTEM

The dynamic as well as the steady state properties of the EGO sensor are of importance to feedback control. To illustrate this feature it is necessary to outline the control strategy itself. We here rely heavily on the pioneering work of Hamburg and Shulman (1980) on this topic and also make further reference to the review by Logothetis (1981). The basic problem is one of characterizing the time delays and attenuations in the total feedback system. The EGO sensor indicates whether the A/F mixture is lean or rich of stoichiometry. If it is lean, for example, the feedback signal to the EFI system (via the microprocessor) can change the A/F supplied to the engine closer to stoichiometry, but this must feed through the engine (and be "processed") before the signal loop back to the EGO is closed, and the exhaust flowing out to the catalytic converter made more nearly stoichiometric, as desired. Through extensive experimentation, Hamburg and Shulman established that the engine ought to be represented by a combination of delay and attenuation elements in the time evolution of the A/F in the closed loop control system. However, in the interest of simplicity, we will here consider it a simple transport delay component only.

Following Logothetis, consider a system with components possessing characteristics as in Figure 3. The A/F delivered to the engine either increases or decreases linearly in time depending upon the signal from a comparator linked to the EGO. When the EGO reads rich, the comparator output link to the EFI causes the A/F to increase. When rich, the opposite occurs. At the initial point in Figure 3, a sensor labelled 1 reads lean and the A/F to the engine is thus made richer. It passes through stoichiometry and a time τ later reaches the EGO, which (here instantaneously) switches to a rich reading. This is conveyed to the comparator and the EFI system, causing a reverse of the A/F engine input. A time τ later, the A/F fed into the engine reaches stoichiometry again, and after a further interval of τ this is detected by the EGO. In this limit cycle mode of operation, therefore, one cycle occurs every 4τ (~ 0.3 to 2 secs). The figure also shows the output of a sensor labelled 2 which has both non-instantaneous response and different relaxation times for the transitions lean to rich (τ_{LR}) and rich to lean (τ_{RL}). It is clear that, were such a sensor to be placed in control, the time the fuel injection system operated in the lean mode would exceed that in the rich mode. A net lean value of A/F would result. On the other hand, if τ_{LR} were to exceed τ_{RL},

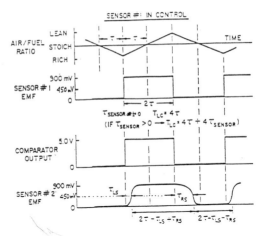

Fig. 3. Limit-Cycle Feedback Control Using Sensor No. 1 See text for discussion of the waveforms

a net rich A/F value would obtain, with possible harmful effects on the catalyst in the long term. Such an eventuality must be avoided by appropriate calibration.

5. ACKNOWLEDGEMENTS

The author is pleased to thank Drs. A. A. Adamczyk, D. R. Hamburg, H. S.Gandhi and E. M. Logothetis for very helpful discussions. It is also a pleasure by this article to acknowledge a long, rewarding, personal and professional relationship with my good friend Ron Bullough.

6. REFERENCES

Brailsford A D and Wynblatt P (1979) *Acta Met.* **27** 489.

Brailsford A D and Logothetis E M (1985) *Sensors and Actuators* **7** 39.

Fleming W J (1980) *Zirconia Oxygen Sensor-An Equivalent Circuit Model* (Warrendale, PA: Society of Automotive Engineers) paper 800020.

Hamburg D R and Shulman M A (1980) *A Closed-Loop A/F Control Model for Internal Combustion Engines* (Warrendale, PA: Society of Automotive Engineers) paper 800826.

Kummer J T (1980) *Prog. Energy Combust. Sci.* **6** 177.

Logothetis E M (1981) *Advances in Ceramics Vol 3,* Edited by A. H. Heuer and L. W. Hobbs; (American Ceramics Society) pp 388-406.

Mizusaki J, Amano K, Yamauchi S and Fueki K (1987) *Solid St. Ionics* **22** 313.

Montreuil C N, Williams S C and Adamczyk A A (1991) *Ford Motor Company Technical Report SR-91-82* (available upon request) - to be published in 1992.

Oh S H and Cavendish J C (1983) *Ind. Eng. Chem. Prod. Res. Dev.* **22** 509.

Rivard J G (1973) *Closed-Loop Electronic Fuel Injection Control of the Internal-Combustion Engine* (New York: Society of Automotive Engineers) paper 730005.

Saji K, Kondo H, Takeuchi T and Igaraski I (1981) *Proceedings of the 1st Sensor Symposium* (The Electrochemical Society) p 103.

Williamson W B Gandhi H S Wynblatt P Truex T J and Ku R C (1980) *AIChE Symposium Ser.* **76** 212.

Wynblatt P and Gjostein N A (1973) *Scripta Met.* **7** 969.

Wynblatt P Dalla Betta R A and Gjostein N A (1975) *The Physical Basis for Heterogeneous Catalysis,* Edited by E Drauglis and R I Jaffee (New York: Plenum) pp. 501-524.

Modelling ceramic superconductors

H Rauh*, A M Stoneham and T C Choy**

AEA Industrial Technology, Harwell Laboratory, Oxfordshire OX11 0RA, England
*Also Department of Materials, University of Oxford, Oxfordshire OX1 3PH, England
**Now Physics Department, Monash University, Clayton, Victoria 3168, Australia

ABSTRACT: Early applications of oxide superconductors are likely to use ceramics rather than single crystals. The performance of these polygranular materials is sensitive to their precise microstructure, which itself depends on processing. We show how different modelling techniques - a discrete mesoscopic approach and effective medium theory - can provide guidance to optimising structures and classifying responses to radiation.

1. INTRODUCTION

The first challenges raised by ceramic superconductors were to basic theory: What is the mechanism? What are the limits on critical temperature? What are the key features of electronic and crystal structure? Technological challenges followed: Where can these systems compete against good conventional materials? What can be achieved with ceramic methods at an industrial scale?

We discuss how theoretical modelling can help these new materials to realise their full potential. For superconductor applications, at least four levels are involved: atomistic modelling to optimise crystal structures and chemical compositions; optimising processing conditions (noting the best phase need not be stable at the natural processing temperature!); optimising microstructure (our concern here); and optimising component design at a macroscopic scale. These levels are not independent. Ideally, macroscopic designs will be based on studies at the microstructural level. If one has to use a polygranular, possibly less than fully dense sample, what are the optimum grain sizes and shapes for a particular property? How can that morphology be achieved? The control of processing depends on what is known of the rate-determining processes studied at an atomic scale. We shall concentrate on how the performance of a given polycrystalline microstructure can best be handled. Our methods also apply to single crystals, where non-uniform stoichiometry, lattice defects or impurities lead to properties which vary from one region to another.

We shall need to know too how much of conventional superconductor theory and conventional oxide technology relates to these new materials still. So far as defect and diffusion processes in ceramic superconductors are concerned, behaviour appears very similar to that of other oxides (Stoneham and Smith 1991). Likewise, the phenomenological London and Ginzburg-Landau theories of superconductivity suffice for most investigations, though we shall have to supplement them through assumptions about weak-links between grains. These assumptions underly current successful attempts to improve superconductor performance and discussions of potential applications (e.g. Doss 1989, Gallop 1990, Alford *et al.* 1991a,b).

We describe two distinct techniques to model polygranular superconductors, with different aims and ranges of validity. For properties like surface resistance (important in microwave applications) or like magnetic susceptibility, a continuum effective medium theory is appropriate: we calculate a "best" average medium to reproduce the actual, complex, system. For critical currents, we turn to discrete mesoscopic modelling, with ensemble averaging over many realisations of the granular structure. This will provide a link to process modelling and texturing too. Our methods are not the only ones used for ceramic superconductors, and we note also the work of Rhyner and Blatter (1989), Vendik *et al.* (1990a,b), and Kim and Torquato (1991).

2. MESOSCOPIC MODELLING: EXPLICIT GRAIN STRUCTURES

In this formulation, there are two separate stages: defining a grain structure and predicting its performance. The first stage generates a realisation of the grain structure. Experimental micrographs of specimens could be used, but these are usually best to check on realism. Instead, a specific granular structure is "grown" using guidelines which mimic actual crystal growth. Often Monte-Carlo techniques are applied, with random nucleation and subsequent development proceeding from a (purely theoretical) mesh. Growth rates which generate aniso-tropic crystals can be allowed, for experiments show major improvements in superconductor performance by texturing; effects of sintering can be modelled too. Indeed, this mesoscopic approach is a route to the optimisation of superconductor processing (Choy *et al.* 1991) and the efficient production of chosen microstructures.

The second stage predicts properties for a series of such structures, and then ensemble averages can be taken. Some properties (like mechanical and high-temperature thermal behaviour) may not involve superconductivity itself. Other properties concentrate on mixtures of superconductors, normal conductors, insulators and voids, both above and below the superconducting transition. Here we shall address critical currents which, in the weakly-linked polygranular structures grown at present, are one or two orders of magnitude short of useful values for technological applications. We shall also look at the energy barriers which impede flux motion. Subtleties can be involved, e.g. the electrical current in the normal and in the superconducting state need not pass through the same grains (as with a one-to-one mixture of YBCO and silver, where the silver component dominates in the normal composite and the YBCO component does once it is superconducting). A similar effect can occur within grains of non-uniform stoichiometry: the current distribution is determined by Kirchhoff's laws for resistances above, and inductances below, the ideal critical temperature T_c. Unusual normal behaviour does not necessarily provide good guidance to the mechanisms of superconductivity!

We have modelled both regular arrays of grains and random microstructures. For simplicity, we shall regard each grain in the polygranular system as an island of homogeneous super-conducting material coupled to others through Josephson-like links. When nearest neighbours only interact, the free energy is controlled by the (temperature-dependent) Josephson critical currents between adjacent grains; it is also governed by the phase differences of the superconductor wave functions (i.e. the Ginzburg-Landau order parameters) across grain boundaries, with additional phase shifts arising from magnetic fields, both self-induced (due to current flow) and externally-applied. The Josephson tunnelling current passing the boundary of a particular grain is given by the derivative of the free energy with respect to the phase of the superconductor wave function of that grain (Josephson 1965).

We seek an iterative determination of the (relative) arrangement of the phases of the super-conductor wave functions which minimises the free energy. This is complicated by the effect of the magnetic fields, which cause frustration, for there are numerous different states almost degenerate in energy; attempts to minimise the free energy lead to oscillations. Examination of the phases shows that flux vortices are present (see Fig. 1 for a regular square array of grains (Shannon and Choy 1991)). The oscillations correspond to movements of the flux. If flux vortices are introduced artificially into a realistic microstructure, moving them through this structure probes the effect of granularity on the free energy. This is shown in Fig. 2 for a polycrystalline microstructure obtained by the discrete mesoscopic approach. With typical values of intergranular coupling, the energies are higher by about 0.1eV when elongated grains lie across the vortex path. This agrees with the qualitative conclusion from studies of regular

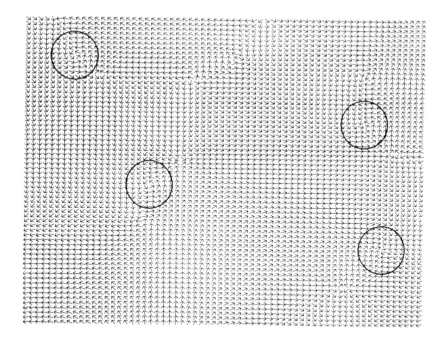

Fig. 1. Phase configuration for a regular square array of grains showing the development of flux vortices (marked); these are associated with circular tunnelling current flows which, in turn, give rise to self-induced magnetic fields.

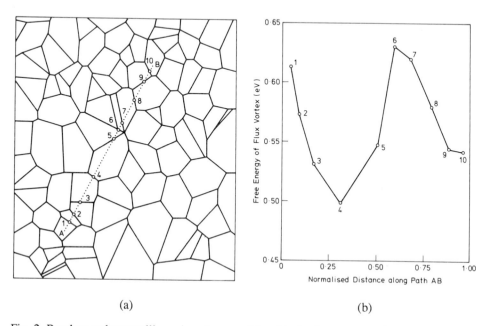

(a)   (b)

Fig. 2. Random polycrystalline microstructure (a), with the free energy of a flux vortex as a function of position across the grains (b). The variations in free energy, of the order of 0.1eV, are a measure for the resistance of the microstructure to (thermally-activated) flux motion.

rectangular arrays of grains, that resistance to flux flow is lower for current flow parallel to the long axes of the grains than for flow across the narrower dimension (Choy *et al.* 1991). In a self-consistent formulation the current distribution would not be imposed, but would itself follow from a calculation starting in the normal state and identifying those superconducting paths which appear as temperature is decreased. Even at this early stage, mesoscopic modelling shows some of the ways that microstructure can influence critical currents.

3. MACROSCOPIC MODELLING: EFFECTIVE MEDIUM THEORY

We turn to a simpler self-consistent approach in which a homogeneous equivalent medium is used to describe a composite made up of several components (perhaps including voids). Such "effective medium" approaches follow the ideas of Maxwell and others (see Landauer 1978); there are links to methods like the coherent potential approximation (Krumhansl 1973) and parallels with distinct methods like percolation theory. The same ideas allow the modelling of the evolution of defect structures under irradiation (Brailsford and Bullough 1981, Rauh *et al.* 1981, 1985, Brailsford 1989, Rauh *et al.* 1992).

The central idea can be illustrated for two conventional dielectrics. Assume there are spherical grains. Calculate the polarisation of a sphere of the first medium in an (as yet undetermined) effective medium in a given static electric field. Repeat the calculation for the second medium embedded in the same effective medium. Assuming specific volume fractions for the two components, calculate the polarisation of the composite. Self-consistency requires this to correspond to the value expected for the effective medium, and the equations yield the average dielectric constant required. In time-dependent (finite frequency) problems the natural description is in terms of scattering by each grain, considered as statistically independent but treated self-consistently. We remark (i) there is a guiding principle that the deviations from the average polarisation are minimised, and that the grains are independent in some sense (e.g. without the long-range magnetic effects which influence the critical current); (ii) the components are treated symmetrically; (iii) we have to use constitutive equations (e.g. Maxwell's equations to describe the fields in the two media); (iv) the volume fractions of the different media are input data, not derived from the calculation; and (v) there is an assumption about grain shape (usually spherical grains). Effective medium theory can be applied to a range of problems, including elasticity, magnetism and current flow.

When we turn to the superconducting ceramics, we need to generalise. First, we shall need the London equations for superconductors, rather than the Maxwell equations, in the limit of weak electric and magnetic fields. Secondly, we shall need to consider at least three components (usually superconducting, normal, and void; we shall not discuss here normal shells covering superconducting grains). The volume fraction of superconducting grains is not determined by composition alone, but also depends on temperature. Since consistency in effective medium theory can be subtle, we take this opportunity to clarify the (correct) equations (10), (11) of Choy and Stoneham (1990a). We must ensure that, as the ideal critical temperature T_c is reached, and hence the London penetration depth λ becomes large, the superconducting inclusions become a part of the effective medium itself. For superconducting grains in their normal state, assigning the average magnetic permeability of the entire composite, $\mu_s = \mu_m$, rather than the magnetic permeability of the normal grains, $\mu_s = \mu_n$, correctly avoids double counting of polarisation terms for normal grains. Both assumptions yield very similar results, identical in the key limits, with modest differences near the percolation threshold where effective medium theory is itself weak.

We shall describe several levels of application of increasing complexity. One key quantity is the volume fraction of superconducting grains. We estimate this using a model which lies outside our effective medium theory. A simple approach is based on an assumed probability distribution of weak-link resistances R at grain boundaries. For a given temperature, the maximum value of R, which still allows phase coherence between adjacent grains, follows from the condition that the Josephson coupling energy associated with critical intergranular tunnelling current flow exceed the energy due to thermal fluctuations; once this criterion is met, phase dependences can be ignored at low fields (Choy and Stoneham 1990a). This is a useful approach, since the main effects of various treatments (irradiation, oxidation or reduction, reaction with moisture or gases) are believed to be dominated by grain boundary changes. These effects can be characterised by changes in the form of the distribution of values of R. With an assumed distribution of weak-link resistances, we can estimate the volume fraction of superconducting grains (strictly the voids should be treated specially, and this is straight-forward).

We are now able to calculate the static electric resistivity (which, of course, falls to zero when a superconducting path extends across the sample, i.e. effectively a percolation limit) and the static magnetic susceptibility (which has a percolation limit too when the magnetic flux is totally expelled). In Fig. 3 we illustrate the temperature dependence of these quantities for a poly-granular YBCO system, as predicted by our effective medium theory, using reasonable working approximations (spherical grains with isotropic material characteristics and a modified Gaussian distribution of weak-link resistances). What do our predictions show? Clearly, the apparent critical temperatures are not the same as the actual value $T_c = 91.6K$ used in our calculations. The temperature obtained from the initial falloff of resistivity differs from that of zero resistivity which, in turn, lies above that of maximum Meissner effect, in agreement with the findings of experiments (Butera 1988, Dubson *et al.* 1988, Härkönen *et al.* 1989). The combination of data from measurements of several properties contains useful information about the intergranular weak-links. Some types of processing (including irradiation, or unintended

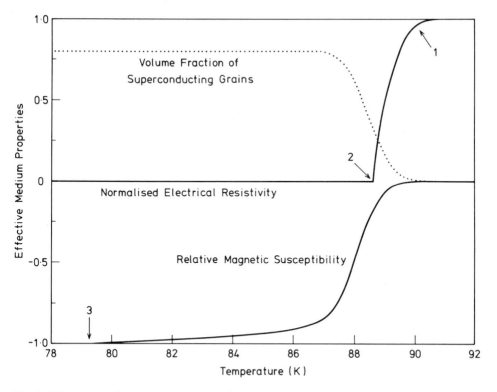

Fig. 3. Effective medium properties: normalised electrical resistivity and relative magnetic susceptibility (together with the volume fraction of superconducting grains) as functions of temperature, predicted for a polygranular YBCO system of 1μm superconducting and normal grains, and 20% porosity. Note the ordering of the temperatures obtained from the initial falloff of resistivity (1), zero resistivity (2), and maximum Meissner effect (3) relative to the ideal critical temperature $T_c = 91.6K$.

atmospheric contamination) mainly affect the R's, and the degradation of performance can usefully be characterised by changes in their distribution. Effective medium theory thus offers a tool for analysing data from radiation experiments and other forms of processing.

Ceramic superconductors have real promise for passive microwave applications, where the critical current poses no limitation (Bybokas and Hammond 1990, Lyons and Withers 1990, Withers *et al.* 1990). Ideal YBCO shows a performance over a significant range of frequencies far superior to that of copper, the material presently in use. Polygranular YBCO fabricated now is less effective, with losses from impurities, from inclusions, and - for thin films - from the substrates, since there is a compromise between substrates best for growth and for dielectric loss. Effective medium theory allows us to examine the effects of inclusions of metallic and dielectric grains (including voids), and the sensitivity to relative grain sizes.

In studying electromagnetic properties of polygranular YBCO systems at finite frequencies, we can identify three regimes, characterised by the respective dominant current types. At lower frequencies, (typically) below about 3GHz, the Meissner current dominates. In this regime, the electric and magnetic polarisabilities are essentially quasi-static, and may be calculated using the Rayleigh scattering approach. For frequencies (typically) up to about 100THz, the eddy current dominates. In that regime, we need the full apparatus of classical Mie scattering theory, suitably modified by London electrodynamics for superconducting grains. The introduction of London electrodynamics entails coupling between the electric and magnetic polarisabilities, a feature probably unique to our present approach. At still higher frequencies, the displacement

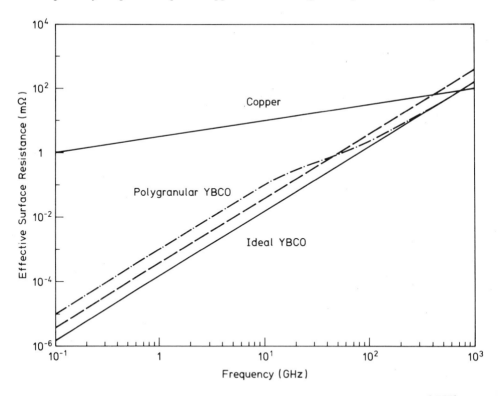

Fig. 4. Effective surface resistance as a function of frequency at a temperature of 77K, predicted for polygranular YBCO systems of 1μm superconducting grains and 1μm (dashed line) or 10μm (dashed-dotted line) granular dielectric inclusions at 10% concentration. Values relating to ideal YBCO and copper (solid lines), obtained from London electrodynamics and classical Maxwell theory, are shown for comparison.

current dominates. Here, the normal dielectric properties prevail (apart from resonances associated with the intergranular weak-links and the superconductor band gap), and super-conductivity loses its significance (Choy and Stoneham 1990b).

The key quantity for microwave applications is the surface resistance. In Fig. 4 we show the frequency dependence of this quantity for polygranular YBCO systems, as predicted by our effective medium theory for a temperature of 77K (again assuming spherical grains with isotropic material characteristics). At this temperature, virtually all YBCO grains will be superconducting, so that their volume fraction is determined by composition alone. Also shown is the frequency dependence of the surface resistance of ideal YBCO and copper, obtained from London electrodynamics and classical Maxwell theory (Choy and Stoneham 1991a,b). The main feature of these results is an increase in loss for polygranular YBCO compared with ideal YBCO, particularly for inclusions of the larger size (though here our prediction of the surface resistance is inadequate above 10GHz, where multiple scattering corrections, not considered in the present effective medium theory formulation, become significant). Polygranular YBCO still shows a much better performance than copper over a useful range of frequencies. Similar results have been obtained with inclusions of metallic rather than dielectric grains, and show a just noticeably smaller loss. It seems unlikely that polygranular YBCO will ever do better than high-quality films, at least if the dielectric loss of the substrates can be reduced. Nevertheless, superconducting ceramics with small grains of high uniformity look very promising.

4. CONCLUSION

The potential of modelling for the optimisation of materials performance is widely realised. We have shown that, even for materials as complex and notoriously difficult in many respects as ceramic superconductors, there is the possibility to improve processing routes and final performances. Even when a full quantitative approach is not appropriate, modelling permits analysis of macroscopic data in terms of microscopic quantities in a way which may be a guide to improvements.

ACKNOWLEDGMENTS

Ron Bullough's initiatives made the superconductivity programme at Harwell possible; his own work in radiation damage introduced modelling techniques which have been transferred to this new field. He has given us tremendous support and scientific insight over the years, and it is a pleasure to express our thanks in this paper. We are grateful too for stimulating discussions with Dr A H Harker, Dr A Hooper, Dr P Schofield, Mr S R Shannon and Dr L W Smith. Some of the present work was funded by the Harwell-Industry Superconducting Ceramics Club and by the UKAEA Programme on Corporate Research.

REFERENCES

Alford N McN, Button T W, Adams M J, Hedges S, Nicholson B and Phillips W A 1991 *Nature* **349** 680
Alford N McN, Button T W, Peterson G E, Smith P A, Davis L E, Penn S J, Lancaster M J, Wu Z and Gallop J C 1991 *IEEE Trans. Magnetics* **27** 1510
Brailsford A D 1989 *Metall. Trans.* **20A** 2583
Brailsford A D and Bullough R 1981 *Phil. Trans. R. Soc. London* **A302** 87
Butera R A 1988 *Phys. Rev.* **B37** 5909
Bybokas J and Hammond B 1990 *Microwave J.* **33**(2) 127
Choy T C, Harding J H, Harker A H, Mulheran P A, Smith L W and Stoneham A M 1991 in *Computer Aided Innovation of New Materials* eds M Doyama, T Suzuki, J Kihara and R Yamamoto (Amsterdam: Elsevier) pp 869-872
Choy T C and Stoneham A M 1990a *J. Phys.: Condens. Matter* **2** 939

Choy T C and Stoneham A M 1990b *J. Phys.: Condens. Matter* **2** 2867
Choy T C and Stoneham A M 1991a *Proc. Fourth Int. Superconducting Applications Convention*, San Diego, California (New York: American Institute of Physics) in press
Choy T C and Stoneham A M 1991b *Proc. R. Soc. London* **A434** 555
Doss J D 1989 *Engineer's Guide to High-Temperature Superconductivity* (New York: John Wiley) ch 4
Dubson M A, Herbert S T, Calabrese J J, Harris D C, Patton B R and Garland J C 1988 *Phys. Rev. Lett.* **60** 1061
Gallop J C 1990 *Supercond. Sci. Technol.* **3** 20
Härkönen K, Tittonen I, Westerholm J and Ullakko K 1989 *Phys. Rev.* **B39** 7251
Josephson B D 1965 *Adv. Phys.* **14** 419
Kim I C and Torquato S 1991 *Phys. Rev.* **A43** 3198
Krumhansl J A 1973 in *Amorphous Magnetism* eds H O Hooper and A M de Graaf (New York: Plenum Press) pp 15-25
Landauer R 1978 in *Electrical Transport and Optical Properties of Inhomogeneous Media* eds J C Garland and D B Tanner (New York: American Institute of Physics) pp 2-45
Lyons W G and Withers R S 1990 *Microwave J.* **33**(11) 85
Rauh H and Bullough R 1985 *Phil. Mag.* **A52** 333
Rauh H, Bullough R and Matthews J R 1992 *Phil. Mag.* **A65** 53
Rauh H, Wood M H and Bullough R 1981 *Phil. Mag.* **A44** 1255
Rhyner J and Blatter G 1989 *Phys. Rev.* **B40** 829
Shannon S R and Choy T C 1991 Presented at the *Fourth Austral. Supercomputer Conf.*, Brisbane, Gold Coast
Stoneham A M and Smith L W 1991 *J. Phys.: Condens. Matter* **3** 225
Vendik O G, Kozyrev A B and Popov A Yu 1990 *Rev. Phys. Appl.* **25** 255
Vendik O G and Popov A Yu 1990 *Mater. Sci. Forum* **62/64** 143
Withers R S, Anderson A C and Oates D E 1990 *Solid St. Technol.* **33**(8) 83

Oxygen precipitation in silicon and intrinsic gettering

R C Newman

Interdisciplinary Research Centre for Semiconductor Materials, The Blackett Laboratory, Imperial College, Prince Consort Road, London SW7 2BZ, UK.

ABSTRACT: Dislocations are preferred sites for the diffusion limited growth of SiO_2 particles in heated silicon but a random distribution of precipitates is produced with an increasing number density as the temperature is reduced.   A three-stage heating schedule leads to a distribution of particles that traps or getters metallic contaminants at sites remote from the surface of a wafer where device structures are fabricated, leading to good manufacturing yields.

1.   INTRODUCTION

The problem of outgassing in vacuum tubes used in early electronic circuits was overcome by evaporating a small charge of barium metal on to the inside of the glass envelope after sealing.   The resulting chemically reactive film removed unwanted gaseous species as they were desorbed from various components. The process was known as gettering.

History has a knack of repeating itself, albeit in an unexpected way.   Surprisingly, the successful fabrication of solid state electronic devices on a wafer of single crystal silicon is also dependent on a gettering process.   Transition metal impurities are introduced as inadvertent contamination during treatments used to form p-n junctions, to grow epitaxial layers or to remove lattice damage following ion implantation, etc. At the high processing temperature the metals are in solution but on cooling, their solubilities decrease and there is formation of internal precipitates (Bullough and Newman 1963).   Metallic precipitates straddling p-n junctions constitute electrical short circuits so that the devices, such as integrated circuits (IC's), do not function (Newman 1982).   Minimising the level of contamination is essential but gettering the residual metal atoms is crucial to obtaining a satisfactory manufacturing yield.

An extensively used method for rendering the metals harmless is known as "intrinsic gettering" and relies on the controlled precipitation of electrically neutral bonded interstitial oxygen impurities present in a concentration $[O_i]_0 \sim 10^{18} cm^{-3}$ in Czochralski (Cz) crystals, which are pulled from a melt contained in a SiO_2 crucible.   Precipitated SiO_2 particles formed during post-growth heat treatments act as a kind of sponge which absorbs the metallic contaminants.   There is still much discussion about the mechanism of the process, which may not be the same for all elements.   Recent work by Gilles et al (1990) implies that Fe atoms segregate to the SiO_2/Si interface so the total surface area and hence the number density of the particles are important parameters.   There must also be a zone denuded of particles at the surface of a wafer where the circuit elements are fabricated.

The purpose of this paper is first to discuss the basic oxygen precipitation process and then to outline the sequence of steps used in intrinsic gettering.

2.   EARLY RESEARCH ON OXYGEN IN SILICON

An infrared (IR) absorption band at 9μm in Cz Si is absent in crystals grown by a zone melting process in which the zone is held in position by surface tension so that there is no contact with a SiO_2 crucible.   The 9μm band, which has a halfwidth $\Delta=34cm^{-1}$ at 300K, is due to the anti-symmetric vibrational mode of $Si-^{16}O_i-Si$ units (Bosomworth et al 1970). For a sample temperature of 4.2K, Δ is reduced to $0.6cm^{-1}$ but the integrated absorption of the line is essentially unchanged.   Values of $[O_i]$ are proportional to the peak absorption coefficient α (cm^{-1}) for a specified value of Δ.   The current calibration factor gives $[O_i]=3.14\alpha \times 10^{17}(cm^{-3})$ at 300K (Baghdadi et al 1989).

Precipitation of isolated O_i atoms results in a reduction of α, as found by Lederhandler and Patel (1957).   They heated Cz Si at 1000°C and showed that the reductions were greater in samples which had been subjected to prior plastic deformation than in as-grown material and it was inferred that dislocations acted as preferred nucleation sites for the growth of SiO_2 particles.   Direct confirmation of the interpretation was obtained by Bullough et al (1960), following earlier related work on the out-diffusion of impurities to a Si surface (Bullough et al 1959).   Briefly, it was recognised that aluminium would displace Si in SiO_2, whereas a similar displacement reaction would not occur with phosphorus, according to the thermodynamic properties of the materials.   Cz crystals doubly-doped with Al and P were grown with [Al]>[P].   During the subsequent heating, SiO_2 precipitates formed and there was concomitant depletion of Al (but not P) from the surrounding region of the Si crystal. Thus the p-type matrix was converted locally to n-type conductivity (Figure 1).   The resulting p-n junctions were revealed by chemical etching and it was demonstrated that dislocations, stacking faults (Figure 2) (Bullough and Newman 1970) and the external surfaces acted as sites for the formation of SiO_2.   The precipitates were also revealed directly by transmission optical microscopy using IR radiation.   The diffused aluminium profiles were analysed by detailed theoretical models taking into account the attraction of the impurities to dislocations due to the elastic size effect interaction (Cottrell 1948, Cottrell and Bilby 1949).   Estimates of the diffusion coefficient D_{Al} obtained in this way were in excellent agreement with those obtained from conventional in-diffusion measurements.

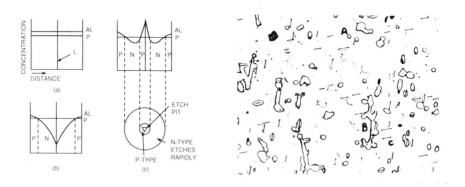

Fig.1.   The formation of p-n and p-n-p structures around a dislocation due to the incorporation and release of Al impurities into SiO_2 precipitates: (a) initial condition, (b) after an anneal at 1200°C and (c) after a second anneal at 1300°C (Newman et al 1965).

Fig.2.   N-type regions around dislocations (A) and stacking faults (B) in Cz Si doped with Al and P (~$10^{17}cm^{-3}$) revealed by etching after an anneal at 1100°C for 16h (Bullough and Newman 1970).

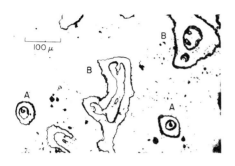

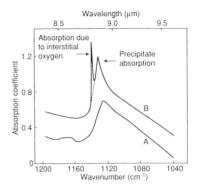

Fig. 3. P-N-P structures in Cz Si revealed by etching around individual dislocations (A) and groups of dislocations (B) following anneals at 1230°C (16h) and then 1275°C (1h) (Newman et al 1965).

Fig. 4. The combined IR absorption from O_i atoms and SiO_2 precipitates measured at 300K (A) and at 4.2K (B) in Cz Si after annealing at 800°C (800h). The traces are displaced vertically for clarity (Livingston et al 1984).

If the anneal temperature was below about 1050°C, SiO_2 precipitates formed at apparently random sites throughout the matrix which was all converted to n-type conductivity, while the precipitate particles were too small to be detected by optical microscopy.

The precipitation process is reversible. Samples given a short second heat treatment at a higher temperature than the first showed p-n-p structures around single dislocations due to the local return of Al and O_i atoms to solution (Figures 1 and 3) (Newman et al 1965).

These results generated several important ideas for future studies. First, measurements of $[O_i]$ following treatments at a given temperature for long times were expected to correspond to the oxygen solubility at that temperature (for a given calibration of the 9μm IR band). An Arrhenius plot should then yield the heat of solution. Early measurements were unsatisfactory because SiO_2 precipitates give rise to broad absorption which underlies the O_i 9μm band (300K) and led to overestimates of the latter absorption. Later measurements with samples held at 4.2K (Livingston et al 1984 and Messoloras et al 1987) allowed the two types of absorption to be separated since the broad absorption from the SiO_2 particles does not sharpen at 4.2K (Figure 4). Second, the concept was established of out-diffusion of impurities to the Si surface, leading to a denuded zone (Bullough et al 1959). The surface concentration should become equal to the solubility after an initial transient and precipitation of the remaining atoms during a subsequent lower temperature heat treatment would be unlikely. Third, the observation of stacking faults implied that there had been generation and aggregation of intrinsic point defects. It was recognised that on average the addition of two O_i atoms to a growing SiO_2 particle would require the emission of one self-interstitial (I-atom) to accommodate the local increase in volume (Bullough and Newman 1970: Gösele 1986). At high temperatures, punching of prismatic dislocations from precipitates sometimes occurred, but at lower temperatures (say 750°C) I-atoms would have to diffuse away from the interface. This conclusion raised the question of whether the growth of SiO_2 would be controlled by O_i diffusion, I-atoms diffusion or by interface reactions. D_{oxy} was not known and no conclusions could be drawn. Fourth, it was clear that a uniform density of SiO_2 would not be produced in crystals with grown-in dislocations and such material would not be suitable for the manufacture of many types of device. This problem was overcome independently since a new method of growing dislocation-free crystals was introduced.

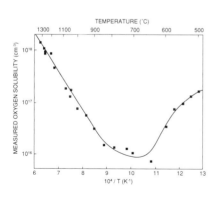

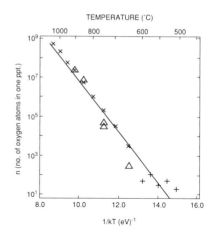

Fig. 5.   The solubility of O_i atoms in Si as measured by IR spectroscopy (4.2K). The upward trend at low temperatures is attributed to the interfacial energy of the small precipitates (see Fig. 6).

Fig. 6.   The average number of O_i atoms per precipitate particle, n, after long heating times deduced from the relationship $n \sim [O_i]_o/N(T,t)$. N(T,t) was determined by SANS and IR measurements (Messoloras et al 1987).

3.   OXYGEN DIFFUSION AND PRECIPITATION PARAMETERS

The profiles of in-diffused $^{18}O_i$ and the out-diffusion of $^{16}O_i$ have been analysed by secondary ion mass spectrography (SIMS) for $T \geq 750°C$. These data, together with measurement of the relaxation of stress-induced dichrosim of the 9µm band (300°C to 1300°C) yielded values of $D_{oxy} = 0.11 \exp(-2.51 \text{ eV}/kT) \text{ cm}^2\text{s}^{-1}$ (Newman 1990).

Measurements were made of the number density N(T,t) and the average size of precipitates, using combinations of chemical etching, transmission electron microscopy, IR scattering, IR(9µm) absorption and small angle neutron scattering (SANS) (Patrick et al 1979. Wada et al 1980, 1982, Livingston et al 1984, Bergholz et al 1989, Messoloras et al 1989, Newman 1991). The SANS data indicated that N(T,t) did not change significantly with heating time, (it decreased slightly), although the behaviour in the earliest stages of precipitate growth was not followed. It was apparent that the total growth process as measured should be described by the theory of Ham (1958), providing O_i diffusion was the rate limiting process. Values of D_{oxy} derived from the modelling, and directly from the SANS measurements, agreed with the SIMS data showing that I-atom emission or surface reactions did not impede the precipitation process. The combined SANS and IR measurements also led to values of the concentration $[O_i]$ that was in equilibrium with the SiO_2 particles which had small radii of curvature after low temperature heat treatments. These solubility data (Figure 5) appear normal down to a temperature $\sim 850°C$ and yield a heat of solution of 1.4 eV. But then there was an apparent increase in the solubility towards a value of $1.6 \times 10^{17} \text{cm}^{-3}$, as the temperature is reduced to 500°C (Messoloras et al 1987). This behaviour was attributed to the surface energy of the particles which incorporate only some twenty O_i atoms at this lowest temperature, according to our analysis (Figure 6). The true solubility should continue to decrease but equilibrium is not achieved because of the low value of D_{oxy}. The measurements explain why it is not possible to nucleate SiO_2 precipitates in Si at low temperatures when the grown-in oxygen content is lower than $\sim 3 \times 10^{17} \text{ cm}^{-3}$. The high temperature edge of the broad minimum in the apparent solubility at 800°C therefore corresponds to the condition where the largest concentration of "reasonably stable" precipitates can be produced. Particles nucleated under these conditions would initially grow more rapidly if the temperature was raised because of the increase in D_{oxy}. However,

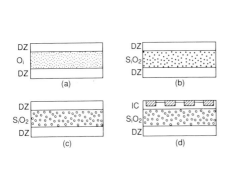

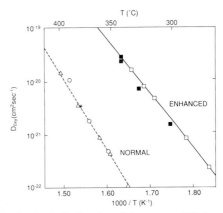

Fig. 7. Heating steps to obtain an intrinsic gettering structure: (a) at 1100°C to form denuded zones (DZ) by out-diffusion of O_i atoms, (b) at T<800°C to nucleate SiO_2 particles in the middle of the wafer and (c) at T>900°C, causing accelerated precipitate growth. IC fabrication is in the denuded zone (d).

Fig. 8. An Arrhenius plot of the oxygen diffusion coefficient D_{oxy} measured by the relaxation of stress-induced dichroism and a second plot showing enhanced diffusion in samples pre-heated in H_2 gas at 900°C for 2h (Newman et al 1991).

at a later stage, Ostwald ripening would occur as the smaller particles redissolved and the larger ones grew even larger.

4. INTRINSIC GETTERING

Three processing steps are necessary to produce the structure required for intrinsic gettering as described in Section1 (Weber and Gilles 1990) (Figure 7). The first step is to heat a wafer at a high temperature to out-diffuse the oxygen to a depth of ~8μm. A suitable temperature would be 1100°C, when the surface concentration $[O_i]$ (equal to the solubility) would be ~$2.5 \times 10^{17} cm^{-3}$, a factor of 4 lower than the grown-in concentration ~$10^{18} cm^{-3}$. The wafer is then heated at T<800°C to nucleate a high density of particles away from the surface, followed by a third heat treatment at T>900°C to cause significant growth of the pre-nucleated particles. Thus, metallic contaminants can be absorbed or gettered by the SiO_2 particles in the middle of the wafer but the top layer is maintained free of both metallic and oxide precipitates. In practice, other heat treatments would be given to the wafer to effect the device fabrication and these treatments will depend on the type of circuit that is being manufactured. The gettering schedule has to be tuned to accommodate all the anneals and no single recipe can be put forward. There is now discussion between device manufactures and vendors of wafers to choose the optimum parameters, including the value of $[O_i]_0$.

5. THE FUTURE

Many integrated circuits are fabricated in lightly doped epitaxial layers grown on Cz Si substrate wafers highly doped with Sb (n-type) or B (p-type). The dopants strongly modify the nucleation behaviour and hence the kinetics of the SiO_2 precipitation in the substrates (Newman 1991). If interfacial dislocations are introduced between the substrate and the epitaxial layer, they will act as preferred nucleation sites for both oxygen and metallic impurities. When epitaxial silicon germanium alloys are used in the fabrication of high frequency (100 GHz) heterojunction bipolar transistors, the problem is more acute because of the lattice misfit.

As we move toward devices with sub-micron features there is a requirement for low temperature processing to minimise effects due to interdiffusion.   Recent work has shown that for T<500°C enhanced O_i diffusion occurs in the presence of hydrogen impurities (Newman et al 1991) (Figure 8).   With the possible use of SiH_4 and Si_2H_6 to grow epitaxial layers, and the now established use of plasma processing, effects due to hydrogen will be of more than academic interest.   There is also recent evidence that small concentrations of nitrogen in Si can affect the oxygen aggregation.   Both H and N impurities can be introduced by heating Si to a high temperature in H_2 and N_2 gases respectively (for review see Newman 1991).

In conclusion, we note that viable gettering procedures have been developed.   There has been an exciting transition whereby oxygen, the impurity that caused problems, has been put to beneficial use in the fabrication of contemporary complex components.   However, research on oxygen in Si is far from complete in spite of the massive international effort during the last thirty six years.

REFERENCES

Baghdadi, A., Bullis, W.M., Croarkin, M.C., Yue-Zhen, Li., Scace, R.I., Series, R.W., Stallhofer, P. and Watanabe, M., 1989, J. Electrochem. Soc. 59, 403

Bergholz, W., Binns, M.J., Booker, G.R., Hutchison, J.C., Kinder, S.H., Messoloras, S., Newman, R.C., Stewart, R.J. and Wilkes, J.G., 1989, Phil. Mag. 59 499

Bosomworth, D.R., Hayes, W., Spray, A.R.L. and Watkins, G.D., 1970, Proc. Roy. Soc. A317 133

Bullough, R. and Newman, R.C., 1963, Prog. Semicond. 7 100

Bullough, R. and Newman, R.C. 1970, Rep. Prog. Phys. 33 101

Bullough, R., Newman, R.C. , and Wakefield, J., 1959 Proc. IEE 106B Suppl. 15 p277

Bullough, R., Newman, R.C., Wakefield, J. and Willis, J.B., 1960 J. Appl. Phys. 31 707

Cottrell, A.H., 1948 Report of Conference on the Strength of Solids, p30 London: Physical Society

Cottrell, A.H. and Bilby, B.A. 1949 Proc. Phys. Soc. 62 49

Gilles, D., Weber, E.R., Haln, S., Monteiro, D.R. and Cho, K., 1990 Semiconductor Silicon 1990 edited by H R Huff, K G Barraclough and J Chikawa, (Electrochem. Soc.: Pennington) 90-7 697

Gösele, U., 1986 Mater. Res. Soc. Symp. Proc. 59 419

Ham, F.S., 1958 J. Phys. Chem. Solids 6 335

Lederhandler, S. and Patel, J.P. 1957 Phys. Rev. 108 239

Livingston, F.M., Messoloras, S., Newman, R.C., Pike, B.C., Stewart, R.J., Binns, M.J., Brown, W.P. and Wilkes, J.G., 1984 J. Phys.C: Solid St. Phys. 17 6253

Messoloras, S., Newman, R.C., Stewart, R. J. and Tucker, J.H., 1987 Semicond. Sci. and Technol. 2 1414

Messoloras, S., Schneider, J.R., Stewart, R.J. and Zulehner, W., 1989 Semicond. Sci. and Technol. 4 340

Newman, R.C., 1982 Rep. Prog. Phys. 45 1163

Newman, R.C. 1990 Proc. 20th Int. Conf. Phys. Semicond. Edited by E M Anastassakis and J D Joannopoulos (World Scientific: Singapore) 1 525

Newman, R C . 1991 Symp. Defects in Silicon 179th Electrochem. Soc, Mtg. Washington, in press

Newman, R.C., Tucker, J.H., Brown, A.R. and McQuaid, S.A., 1991 J. Appl. Phys. in press

Newman, R.C., Wakefield, J. and Willis, J.B., 1965, Solid St. Electron. 8 180

Patrick, W., Hearn, E., Westdorp, W. and Bohg, A. 1979 J.Appl.Phys. 50 7156

Wada, K., Inoue, N. and Kohra, K. 1980 J.Crystal Growth 49 749

Wada, K., Nakanishi, H., Takaoko, T. and Inoue, N. 1982 J.Crystal Growth 57 535

Weber, E. and Gilles, D., 1990 Semiconductor Silicon 1990 edited by HR Huff, K G Barraclough and J Chikawa (Electrochem. Soc.: Pennington) 90-7 585

Modelling the response of semiconducting oxides as gas sensors

D E Williams

Dept of Chemistry, University College London, 20 Gordon St, London WC1H 0AJ, U.K.

ABSTRACT: The atmosphere-dependent variation of conductivity of ceramic oxides has resulted in the industrial development of small, rugged and inexpensive devices for a variety of gas measurement tasks. Improvement of these devices requires a thorough knowledge of the response mechanism; modelling plays a central part in the development of such understanding. This article outlines current ideas, emphasising the use of simple physical models to rationalise a wide range of sometimes apparently disparate observations.

1.INTRODUCTION

Semiconducting oxides at elevated temperature show conductivity changes in response to changes in composition of the gaseous atmosphere, as is well known (Moseley and Tofield, 1987). This property of ceramic oxides opens the possibility of small, rugged and inexpensive devices for a variety of gas measurement tasks: as a consequence, a large research and development effort has been spawned. A number of physical phenomena are involved in the transduction of a change in the composition of a gaseous atmosphere into a change in conductance of a ceramic oxide. These processes are illustrated schematically in Figure 1. An examination of each of the three general classes of process involved, namely those in the gas phase, on the surface and in the bulk, reveals just how complex can be the relationship between observable (conductivity change) and the desired analysis (bulk gas composition or equilibrium gas composition, for example). It is in unravelling this complexity that modelling is of great importance.

Firstly, the gas composition: in the bulk, this may be at equilibrium, but more generally is not. A small concentration of carbon monoxide in air at room temperature is not at equilibrium; neither is the complex gas mixture ejected from the exhaust valve of an internal combustion engine. At the sensor surface, reactions between the components of the gas mixture may be catalysed, and the mixture may attain an equilibrium composition there; equally, it may not, although the composition is likely to be different from that in the bulk gas phase. When, as is usual, the sensor is fabricated as a fine-grained porous body, the composition of the gas mixture and therefore the local electrical conductivity of the sensor material, might, as a consequence, vary throughout the sensor. A response results which is not necessarily simply related to the composition of the gas in the bulk far away from the sensor surface (Brailsford and Logothetis 1985, Williams 1987).

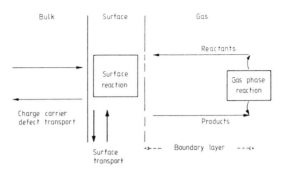

Figure 1   Illustration of physical phenomena involved in the transduction of a change in the composition of a gaseous atmosphere into a change in conductance of a ceramic oxide. Surface reactions may be between adsorbed species and defects in the solid, or between different adsorbed species, or between gas molecules and adsorbed species or any or all of these together. The surface reaction may involve a catalyst (Williams 1987: reproduced with permission).

Secondly, the equilibrium between the oxide bulk and surface: electronic equilibrium between surface and bulk should be established, but equilibrium of ionic species may not. Oxygen vacancies, for example, might be mobile, in which case their concentration in the immediate vicinity of the interface would be at equilibrium with that in the bulk solid; equally, such vacancies may not be mobile, in which case the surface and bulk of the solid would not be at equilibrium. If the vacancy mobility were small but non-zero, there might be slow drifts of the response with time (Göpel, 1989).

Thirdly, the measured conductance: this reflects both the concentration and mobility of the charge carriers in the solid phase. If the sensor is in the form of a porous body, then a key element to consider is the transport of charge carriers across the junctions between grains.

Finally, there are the reactions at the interface between gas and solid. These reactions couple changes in gas composition at the interface to changes in charge carrier concentration. If the solid side of the interface is not at equilibrium with the gas side, then the problem arises as to the precise chemistry of the surface reactions and which of them does the coupling to charge carrier concentration. Further, in the absence of equilibrium, the question arises as to the surface concentrations of the important intermediates and, indeed, what these are.

The two most common types of gas sensor based upon conductivity changes in oxide materials exemplify two extremes. First there is the sensor type based upon materials such as TiO_2, in which both ionic and electronic species are at equilibrium between bulk and surface zones of the solid and the species present on the surface of the solid are in equilibrium with the local oxygen partial pressure. At high enough temperature, all oxide materials behave in this way, and the variation of conductivity with oxygen partial pressure reflects the dominant defect equilibria in the solid (Kofstad, 1972).

The other extreme of sensor material is exemplified by SnO_2, where, at moderate temperatures (less than about 800K) there is no ionic equilibrium between bulk and surface of the solid phase. Furthermore, the sensor surface is not at equilibrium with the local gas composition, and the gas itself is a non-equilibrium mixture. Metal oxides of this type, such as tin dioxide and zinc oxide, fabricated either in the form of thin (~100nm) films or as thicker porous bodies, show an electrical conductivity at temperatures around 600K which is very sensitive to the presence of traces (ppm level) of reactive gases (hydrocarbons, hydrogen, carbon monoxide, methane, ammonia, oxides of sulphur and nitrogen, chlorine, hydrogen sulphide...) in air. This phenomenon has been exploited for many years in warning devices. The majority of commercial elements utilize porous, thick (~100m) films of tin dioxide. For these non-equilibrium devices, considerations of gas transport and reaction within the porous mass of the sensor body must form a central part of the interpretation of the behaviour, since these devices appear necessarily to cause combustion of the gas mixtures for whose analysis they are applied. Thus, the local conductivity of a part of the sensor will reflect the local gas composition, which will not be the same as the bulk gas composition and will not be the same everywhere inside the porous sensor mass. The observed response is then dependent upon the sensor geometry and the diffusion and combustion rates of the gas: very marked effects can be observed, especially in devices in which have been incorporated combustion catalysts such as precious metals (McAleer et al, 1988). Simple reaction-diffusion models can be set up to illustrate the variation of gas composition with position within the porous sensor body, the local conductivity can then be related to the local gas composition and finally, the conductance of the sensor calculated given the geometry of sensor and the geometry and position of the electrodes (Williams 1987, Jain et al 1990).

2.A SURVEY OF RESPONSES OF OXIDES TO NON-EQUILIBRIUM GAS MIXTURES

It turns out that the phenomenon of a conductance response to the presence in air of a trace concentration of a reactive gas, is common to many different oxide materials (Moseley et al 1991). Indeed, it seems that, provided the conductivity in air in the temperature range approximately 500K to 800K is neither too high, in which case the carrier density is too large for small changes to be detected, or too low, in which case the conductance may be too low to be measured reliably, then a conductance change will be observed if, into the air, is introduced a small concentration (1000vpm, say) of a reactive gas such as hydrogen, carbon monoxide, hydrocarbons or ammonia, to name but a few. A simple general rule can be adumbrated, which describes the general patterns of response that can be observed: gases can clearly be classified into 'oxidising' and 'reducing', and oxides into 'p-type' and 'n-type'. Thus, to an oxidising gas (including increase of partial pressure of oxygen itself), the response of an n-type oxide is either a conductivity decrease or no response and of a p-type oxide a conductivity increase or no response; to a reducing gas (including decrease of oxygen partial pressure) the responses, if any, are opposite in sign. There are, of course, exceptions to this rule, and the insight which develops from a consideration of the exceptions is described later. Figure 2 illustrates one of the exceptions.

The fact that there is a rather general pattern of behaviour, implies that different oxide materials respond to the presence of trace reactive gases in a rather similar way. Thus, there does not appear to be a large range of different patterns of response, such as might be obtained if each oxide had some unique response dependent on very specific aspects of the surface chemistry. The implication is that the major response of an oxide material to a range of gases comes from the reaction of these gases with a single surface species which controls charge carrier concentrations in the vicinity of the gas-solid interface. Most studies point to this species being an adsorbed ionised oxygen species such as O_2^- or O^- (Moseley and

Williams 1991), and various kinetic schemes can be written to rationalise the dependence of surface coverage of this species and its connection with conductivity variations. The whole question of the surface chemistry of oxides in non-equilibrium gas atmospheres, and the coupling of the chemistry to changes in conductivity, is one which has not been much addressed, but which is in fact central to understanding of the behaviour of sensors of this type.

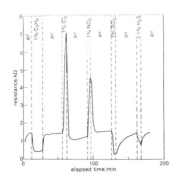 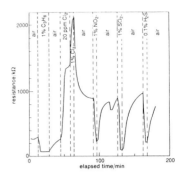

Figure 2   Response at 500C of the resistance of a porous pellet of oxide material to both oxidising and reducing gases in air illustrating n-type behaviour: (a) $BaSnO_3$; and mixed behaviour dependent upon the nature of the gas present : (b) $Ba_6FeNb_9O_{30}$; taken with permission from Moseley et al (1991).

3.A GENERAL DESCRIPTION OF MODELS FOR THE CONDUCTANCE OF OXIDES IN WHICH THE SURFACE AND BULK ARE NOT IN IONIC EQUILIBRIUM

Species adsorbed at the solid-gas interface are assumed to be in electronic equilibrium with the solid; thus, the basic idea is that oxygen, the dominant species adsorbed at the gas-solid interface, abstracts electrons from the solid to form a surface oxygen ion. Adsorbed oxygen can be thought of as a surface trap state for electrons. In the case of an n-type oxide, the charge carrier density at the interface is thereby diminished and a potential barrier develops at the interface. The adsorption rate diminishes with increasing coverage of oxygen, and the coverage saturates at a low value. In the case of p-type oxides, adsorbed oxygen acts as a surface acceptor state, abstracting electrons from the valence band and hence giving rise to an increase in the charge carrier (hole) concentration at the interface. In contrast to the situation with n-type oxides, the surface coverage could proceed to a rather large value before saturating. The key point is that the charge carrier concentration in the vicinity of the interface is expected to be dependent upon the surface coverage of oxygen species. The assumption is then that, in the presence of a reactive gas, a surface-catalysed combustion occurs, and that the surface coverage of oxygen species is then determined by a kinetic steady state: the change in charge carrier concentration in the vicinity of the interface, and consequently the change in conductance, reflects the resultant change in surface oxygen concentration.

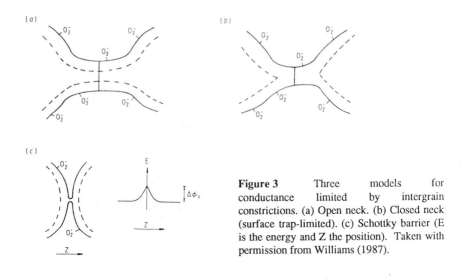

Figure 3 Three models for conductance limited by intergrain constrictions. (a) Open neck. (b) Closed neck (surface trap-limited). (c) Schottky barrier (E is the energy and Z the position). Taken with permission from Williams (1987).

The way in which changes in the charge carrier concentration in the vicinity of the interface result in changes in measured conductance can only be appreciated by constructing idealised models of the physical form of the solid. In the case of devices which can be modelled as ideal, structure-free, thin layers, the connection is direct: if the depletion layer thickness for majority carriers (for n-type materials; accumulation layer thickness for p-type) represents a substantial fraction of the layer thickness, then, clearly, the measured conductance of the layer, parallel to the interface, would directly reflect changes in the charge carrier density at the interface. The connection is more subtle in the case of real devices. Where charge transport requires transport between different grains of the solid, particularly in the case where the sensor is in the form of a highly porous mass, then the effects are expected to be dependent upon the grain size of the solid and upon the dimension of the conducting paths between the grains. Figure 3 (Beekmans 1978) illustrates three idealisations of intergrain contacts for an n-type oxide: in the first, the depletion layer does not extend completely across the intergranular neck, so the conductance across this junction would change with changes in the depletion layer thickness; in the second, the depletion layer extends across the junction so the conductance would be controlled by the excitation of electrons from the surface traps (surface-trap limited conductance); in the third, the junction is short and narrow and so behaves as a kind of potential barrier between the grains, which charge carriers have to cross (Schottky barrier limited case). Some results in the literature (McAleer et al 1987) have indicated a Schottky barrier-limited conductance for tin dioxide sensors; in others (Jones et al 1984), the interpretation has been based upon a surface trap-limited conductance. As indicated in the following and implied by the discussion above, it is quite likely that the effects are sensitively dependent upon the microstructure and particularly upon the grain size of the material. In this connection, another possible subtlety, necessary for the interpretation of the conductance behaviour of porous bodies, should be mentioned. If the electrical resistance of certain types of intergrain contact (the Schottky barrier type, for example) were much greater than that of other types ('necks', say), then it becomes interesting to consider the consequences of the following theoretical argument. In a three-dimensional mass, in which each crystallite is connected to several neighbours by contacts which may be either 'necks' or 'barriers', there will exist a percolation limit pertaining to the proportion of contacts which are 'barriers'. If this proportion is higher than the percolation limit, then current carriers must necessarily cross a 'barrier' somewhere. With a proportion of 'barriers' greater than the limit, an effective domain size can be defined, which is a function of the proportion of 'barriers'. Clearly, this effective domain size, and its stability with time would be expected to have an important effect upon such practical questions as the repeatability of characteristics from one sensor to another and the stability of the response with time: changes could be consequent upon changes in microstructure caused for example by thermal expansion and contraction, or further sintering at the operating temperature, or by reactions with components of the gas atmosphere, moisture particularly. Furthermore, in a porous sensor body, there is a second sort of percolation problem, that of the access of the gas phase to all parts of the interior of the body : not all parts of the interior of a porous mass may be easily accessible to a gas. The effect of closed porosity, or of extremely fine porosity, might also be to define domains, such that the gas sensitivity of conductance arises from connections between domains.

In the case of p-type oxides, in a conventional description, the adsorbed oxygen acts as a surface acceptor state, abstracting electrons from the valence band and hence giving rise to an increase in the charge carrier (hole) concentration at the interface. The grain junctions therefore would have lower resistivity than the bulk of the material, although they could still represent a significant element in the resistance of the solid because they are so narrow. Now, in contrast to n-type materials, any decrease in the surface coverage of oxygen species would lead to a decrease in the charge carrier concentration in the grain junctions and hence to an increase in the resistance of the material. In contrast to the n-type materials, also, it would seem that no barrier to charge carrier transport can be developed at grain junctions as a consequence of the oxygen adsorption.

The discussion above has emphasised the importance of connections between crystallites or domains of crystallites, and has indirectly thereby indicated the potential complexity of modelling. It is useful, however, to consider the consequences of a simple limiting case (Williams and Moseley 1991): if all the crystallites were equally accessible to the gas, if the crystallite size were small enough and if the bulk donor density were low enough, then as a consequence of adsorption of oxygen the crystallites might be completely depleted of conduction electrons and a surface trap-limited conductivity might then ensue: n- and p-type oxides could then be treated within a common framework since in neither case would there be a potential barrier to charge transport across the grain junctions. The porous mass, in the absence of potential barriers limiting transport between domains of the solid, can be considered as the equivalent non-porous solid, in which the acceptor state density per unit volume is variable: the acceptor state density replaces the oxygen partial pressure as the independent variable controlling the behaviour of the effective medium, and in particular the transition from one type of behaviour (n-type) to the opposite type (p-type). The apparent acceptor state density per unit volume depends, as noted above, upon the actual surface acceptor state density of the porous mass, which in turn is dependent upon the non-equilibrium chemistry linking it to the gas composition in the vicinity of the interface. The consequence of this simplification is that a rationalisation of the behaviour of oxide sensors can be obtained, including the apparent exceptions to the general rule set out earlier.

Figure 4 illustrates schematically the result of the calculation (Williams and Moseley 1991). Given that the effect of the presence of the gas is assumed to be a change in the surface acceptor state density, with reducing gases giving a decrease and oxidising gases an increase in its value, then it is clear that, with a sufficiently high bulk donor density, purely n-type behaviour is expected (curve 1: conductivity increase in the presence of a reducing gas causing a decrease in surface acceptor state density). If the bulk donor density is low enough (curve 3), then purely p-type behaviour is expected (conductivity decrease in the presence of a reducing gas) since there is evidently a limit to the extent to which the surface acceptor state density can be lowered when the reducing gas is only a minor component of the atmosphere. In the intermediate case (curve 2) the observed conductivity could first decrease and then increase with increasing concentration of a reducing gas, or one gas being a mild reducing agent might cause conductivity decreases only over the range of concentration studied whereas another, more powerful reducing agent might cause conductivity to go through a minimum and then rise over the same range of concentration. The effect of oxidising gases can be similarly rationalised with the small difference that such gases are expected to cause an increase in the surface acceptor state density. It can immediately be seen that the calculation provides a neat rationalisation of the whole range of the observations. It also provokes another insight: if the donor and acceptor densities are such that the normal condition of the material is around the minimum in the curve of conductance against surface acceptor state density, then the response to gases reacting with surface oxygen species will be minimum. In that case, other influences upon the conductivity, such as specific complexing reactions of gases with surface sites, might become observable. It is possible that some of the specific effects of ammonia upon conductance of certain oxides (Williams 1987, Moseley and Williams 1990) could be explained in this way.

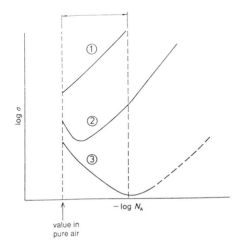

Figure 4 Schematic illustration of the variation of conductance, σ, of a porous sensor body, with variation in the surface acceptor state density, N_A, assuming a surface trap-limited conductance. The three curves show the expected variation with progressively decreasing bulk donor density, N_D : $N_{D,1} > N_{D,2} > N_{D,3}$. The region indicated by the arrows delineates schematically the accessible range of surface acceptor state density when a reducing gas is present as a minor component in air; taken with permission from Williams and Moseley (1991).

4.THE IMPORTANCE OF SURFACE CHEMISTRY

The discussion above has nothing to say about specific properties of materials, apart from the general distinction between n- and p-type. All oxides are expected to adsorb and dissociate oxygen to some extent; so the implication is that, provided that a surface reaction with the target species can generate a departure from the equilibrium oxygen surface coverage, and provided that the bulk resistivity is in a range such that the resultant changes in resistance can be detected, then a gas-dependent conductance change will be observed. For a surface reaction to give an electrical response, there must be some coupling with the charge carrier concentration. The surface chemistry is the critical link between the target measureand and the observable result, yet there has been surprisingly little study of this aspect of the response mechanism. Such studies as have been made suggest that rather minor branches of the surface reaction, for example, that leading to the production of formaldehyde on SnO_2 in the presence of trace carbon monoxide in wet air (Kohl 1989), can have the dominant effect upon the measured conductivity change. In a qualitative way, it can be argued that the relative differences in magnitude of response of different materials to the same concentration of a particular gas, and the relative differences in response of a given material to the same concentration of different gases, arise because the rate of the surface catalysed combustion reactions, for example, may be very different for different gases and different oxides, and indeed that certain surface structures might specifically enhance or retard certain reactions. This is a part of the problem of the behaviour of these devices that modelling might help to address in the future.

5.REFERENCES

Beekmans N M 1978 *J Chem Soc Faraday Trans I* **74** 31-45
Brailsford A D and Logothetis E M 1985 *Sensors and Actuators* **7** 39-67
Göpel W 1989 *Sensors and Actuators* **16** 167-93
Kofstad P 1972 *Non-stoichiometry, Diffusion and Electrical Conductivity in Primary Metal Oxides* (New York: Wiley-Interscience)
Kohl D 1989 *Sensors and Actuators* **18** 71-116
Jain U, Williams D E, Harker A H and Stoneham A M 1990 *Sensors and Actuators B* **2** 111-114
Jones A, Jones T A, Mann B and Firth J G 1984 *Sensors and Actuators* **5** 75
McAleer J F, Moseley P T, Norris J O W and Williams D E 1987 *J Chem Soc Faraday Trans I* **83** 1323-46
McAleer J F, Moseley P T, Norris J O W, Williams D E and Tofield B C 1988 *J Chem Soc Faraday Trans I* **84** 441-57
Moseley P T, Stoneham A M and Williams D E 1991 *Techniques and Mechanisms in Gas Sensing* ed P T Moseley, J O W Norris and D E Williams (Bristol and Philadelphia: Adam Hilger)
Moseley P T and Tofield B C (Ed) 1987 *Solid State Gas Sensors* (Bristol and Philadelphia: Adam Hilger)
Moseley P T and Williams D E 1990 *Sensors and Actuators* **B1** 113
Moseley P T and Williams D E 1991 *Techniques and Mechanisms in Gas Sensing* ed P T Moseley, J O W Norris and D E Williams (Bristol and Philadelphia: Adam Hilger)
Williams D E 1987 *Solid State Gas Sensors* ed P T Moseley and B C Tofield (Bristol and Philadelphia: Adam Hilger)
Williams D E and Moseley P T 1991 *J Materials Chem* **1** 809-814

Subject Index

Author Index